Table of Contents

Introduction

One of the most critical aspects of transforming the nation's public safety answering points (PSAPs) from today's legacy 911 technology to Next Generation 911 (NG911) is adherence to a common set of standards. Development and adoption of international standards is key to achieving 911 interoperability across multiple local, regional, state, and national public safety jurisdictions, and beyond into the global emergency communications environment. Based on conceptual definitions dating from 2000, development began on NG911 standards in 2003 when the National Emergency Number Association (NENA) initiated technical requirements and definition work on core Internet Protocol (IP) functionality and architecture.

Beyond the walls of the 911 PSAPs, the consistent observance of standards is essential in accomplishing seamless transmission of data from the caller to 911, and on to emergency responders. As PSAPs expand the forms of data they receive and transmit to each other, and as emergency responders migrate to a broadband network (e.g., FirstNet), it is essential that standards are established and consistently adopted.

A variety of standards already exist, and many are actively under development. However, there is limited coordination across the broad NG911 community regarding what completed standards are available, what standards overlap, and what standards still need to be established. The National 911 Program, led by the United States (U.S.) Department of Transportation (USDOT), National Highway Traffic Safety Administration (NHTSA), has compiled this list of standards activities related to NG911. The standards development organizations (SDOs) mentioned herein were given the opportunity to vet the contents of this document, to assess the status of specific standards. This is a living document, and the National 911 Program will publish,[1] monitor, support, and promote the activities of SDOs in establishing a comprehensive set of standards for NG911.

Input from the standards community and NG911 stakeholders at large is encouraged and appreciated. The National 911 Program can be reached at (202) 366-3485 or via email at: nhtsa.national911@dot.gov.

---

[1] Available through the National 911 Program at: http://www.911.gov.

What Is a Standard?

The International Organization for Standardization (ISO)/International Electrotechnical Commission (IEC) Guide 2:2004, definition 3.2, defines a standard as a[2]—

> *document, established by consensus and approved by a recognized body, that provides, for common and repeated use, rules, guidelines or characteristics for activities or their results, aimed at the achievement of the optimum degree of order in a given context*

Standards affect the daily lives of everyone across the nation. From the most mundane aspects of life (e.g., electrical cords and wall sockets) to potentially life and death situations (e.g., the concentration of ingredients in generic medications), standards guide the quality, safety, and security of products or processes. Standards are widely used in all areas throughout the U.S. government and public and private sectors.

Standards can be *voluntary*—by themselves imposing no requirement regarding use—or *mandatory*. Generally, a mandatory standard is published as part of a code, rule, or regulation by a regulatory government body and imposes an obligation on specified parties to conform to it. However, the distinction between these two categories may be lost when voluntary consensus standards are referenced in government regulations, effectively making them mandatory standards.[3] Most standards are **voluntary, consensus-based**, and **open**:[4]

- Voluntary—Use of the standard is not mandated by law
- Consensus-based—Published standards have attained general agreement through cooperation and compromise in a process that is inclusive of all interested parties
- Open—Standards are not proprietary and are available for anyone to use

A standard may be or contain intellectual property such as patents, and the intellectual property rights (IPR) may still be held by a company. The American National Standards Institute (ANSI) essential elements state this about patents in ANSI standards:

---

[2]ISO, *ISO/IEC Directives, Part 2: Rules for the structure and drafting of International Standards.* Available at: http://isotc.iso.org/livelink/livelink?func=ll&objId=4230456&objAction=browse&sort=subtype (last accessed February 24, 2016).

[3] Standards.gov, *What Are Standards?* Available at: http://www.nist.gov/standardsgov/definestandards.cfm (last accessed March 15, 2016).

[4] RITA Intelligent Transport Systems, *What Are Standards?* Available at: http://www.standards.its.dot.gov/LearnAboutStandards/ITSStandardsBackground (last accessed March 15, 2016).

The ASD shall receive from the patent holder or a party authorized to make assurances on its behalf, in written or electronic form, either:

a) assurance in the form of a general disclaimer to the effect that such party does not hold and does not currently intend holding any essential patent claim(s); or

b) assurance that a license to such essential patent claim(s) will be made available to applicants desiring to utilize the license for the purpose of implementing the standard either:

> *i) under reasonable terms and conditions that are demonstrably free of any unfair discrimination; or*

> *ii) without compensation and under reasonable terms and conditions that are demonstrably free of any unfair discrimination.* [5]

What Are Best Practices?

Typically less formal than standards, best practices are methods or techniques that have been identified as the most effective, efficient, and practical means to achieve an objective. Based on a repeatable process, best practices often emerge as the result of generally accepted principles followed by many individuals, groups, or organizations, which have been established over time. Best practices often supplement the standards process and act as common guidelines for policies and operations.

Stakeholders

Stakeholders in standardization encompass all groups that have an interest in a particular standard because those groups are likely to be most affected by changes and, therefore, want to contribute to the development process. NG911 stakeholders are members of a broad and diverse community of users who generally can be categorized as follows:

- 911 and public safety agencies and authorities
- Vendor community (including hardware and software) and related industries
- Technology, services, and consulting industries
- SDOs and standards setting organizations (SSOs)
- Consumer, research, academic, and consortia communities
- Telematics, third-party call centers, Internet, infrastructure, wireline, and wireless service providers
- Transportation agencies
- Local, state, and federal governments

---

[5] ANSI Essential Requirements: Due process requirements for American National Standards, January 2015, Page 10. As viewed at: http://publicaa.ansi.org/sites/apdl/Documents/Standards%20Activities/American%20National%20Standards/Procedures,%20Guides,%20and%20Forms/2015_ANSI_Essential_Requirements.pdf (last accessed February 24, 2016).

- Regulatory agencies and public utility commissions
- Professional and trade associations
- The public at large[6]

Standards Organizations

Standards organizations are bodies, organizations, and institutions whose focus is developing and maintaining standards in the interest of a user community. These organizations can be governmental, quasi-governmental, and non-governmental.[7] Typically, their mandate is geographically oriented—international, regional, or national. Organizations that establish, review, and maintain standards are considered to be SDOs,[8] although consortia are sometimes differentiated as SSOs. Generally speaking, SDOs and SSOs consistently adhere to a set of requirements or procedures that govern the standards development process.

How Are Standards Developed?

At the heart of the U.S. standards system are voluntary standards that arise from a formal, coordinated, consensus-based, and open process. Developed by subject matter experts from both the public and private sectors, the voluntary process is open to all affected parties and relies on cooperation and compromise among a diverse range of stakeholders. Organizations also work together to develop joint standards, which forge relationships and allow for a collaborative and cooperative effort. Joint standards will be especially important with respect to the synergistic environment of emergency communications, such as the environment shared by the Nationwide Public Safety Broadband Network (NPSBN) and NG911.

Although the development process may vary to some extent from organization to organization, fundamentally each organization has an established set of formally documented procedures for initiating, developing, reviewing, approving, and maintaining standards. As an example, the following diagram illustrates the USDOT Research and Innovative Technology Administration (RITA) Intelligent Transportation Systems (ITS) standards development process:[9]

---

[6] Although it is generally accepted that the public is an NG911 stakeholder (as the primary 911 call originator), typically, any involvement with the standards process occurs only when they participate as part of another stakeholder group.

[7] Quasi- and non-governmental standards organizations are often non-profit organizations.

[8] Standards Development Organization or Standard Developing Organization.

[9] Research and Innovative Technology Administration (RITA) Intelligent Transportation Systems (ITS), *Standards Development Process.* http://www.standards.its.dot.gov/LearnAboutStandards/StandardsDevelopment (last accessed February 24, 2016).

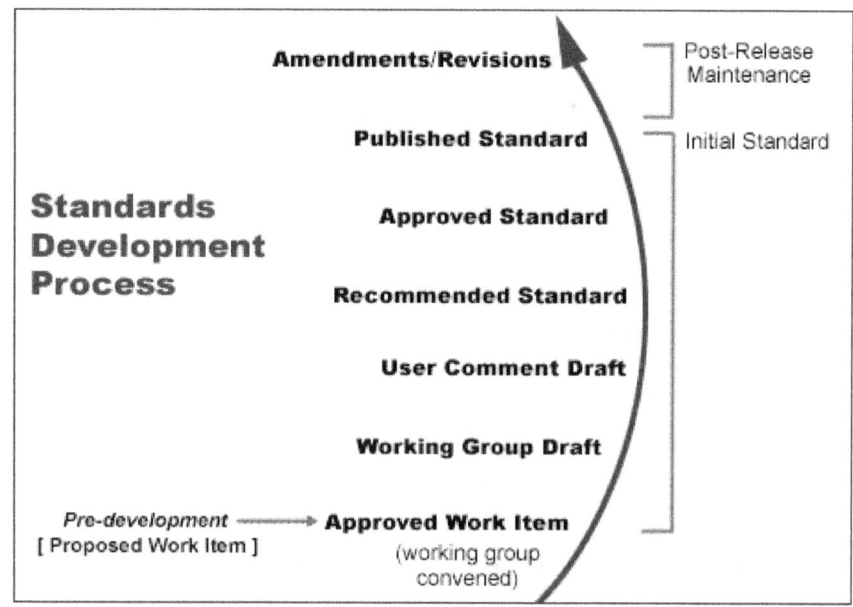

The Institute of Electrical and Electronics Engineers (IEEE) emphasizes that standards "are 'living documents', which may initially be published and iteratively modified, corrected, adjusted and/or updated based on market conditions and other factors."[10] Given that standards development is an iterative process, often there are procedures for publishing draft and/or interim documents at different stages in the process prior to formal approval. Once approved, various factors can render standards outdated, including technological advancements and new or revised requirements. For this reason, the majority of standards require periodic review and, potentially, revision. As a general rule, organizations such as ANSI and ISO assert that standards should be reviewed at intervals of not more than five years.[11]

What Is Standards Accreditation?

Typically, process accreditation bodies do not develop standards but instead provide accreditation services for the purpose of assessing and certifying the standards development process of other SDOs. For example, ANSI facilitates development of American National Standards (ANS) by accrediting the procedures of SDOs. Accreditation by ANSI signifies that the procedures used by the standards body, in connection with the development of ANS, meet the Institute's essential requirements for openness, balance, consensus, and due process.[12] Given the voluntary nature of standards, SDOs are not mandated

---

[10] IEEE Volunteer Training Program, *How are Standards Made?* Available at:
http://standards.ieee.org/develop/process.html (last accessed February 24, 2016).
[11] ANSI Essential Requirements: Due process requirements for American National Standards, January 2015, page 20. As viewed at:
http://publicaa.ansi.org/sites/apdl/Documents/Standards%20Activities/American%20National%20Standards/Procedures,%20Guides,%20and%20Forms/2015_ANSI_Essential_Requirements.pdf (last accessed February 24, 2016).
[12] ANSI Standards Activities, *Domestic Programs (American National Standards) Overview.* Available at:
http://www.ansi.org/standards_activities/domestic_programs/overview.aspx (last accessed February 24, 2016).

to attain accreditation. However, accreditation does demonstrate adherence and conformity with a formal and recognized standards development process. Given the expense and time involved, not all SDOs pursue accreditation, although they are still likely to adhere to a similarly rigorous standards development process.

Types of Standards

In an effort to organize the numerous standards that are of interest and applicability to the NG911 community, this document groups standards into the following six categories:

- **Product Standard**—Describes the expectations and minimum requirements for a particular product, typically in the context of a specific use. Product standards would most often be reflected in descriptions of hardware, software, and other technology solutions.
- **Interface Standard**—Describes the requirements for connecting two or more systems, or technologies, to one another. User interface standards would describe the interconnection between a human and a machine.
- **Data Standard**—Describes the definition, format, layout, and other characteristics of data stored within a system or shared across systems. Data standards help to ensure the seamless exchange of data between disparate systems and permit a common understanding to interpret and use data consistently.
- **Test Standard**—Describes the test methodologies, processes, and other requirements associated with determining the performance or fitness of a particular product.
- **Performance Standard**—Describes how a product or service should function, often in terms of quality, quantity, or timeliness.
- **Operational Standard**—Describes how a function or business process should occur, setting minimum requirements for performance or delivery. Operational standards could include standard operating procedures (SOPs), training guidelines, and policies.

The first three categories (product, interface, and data) are primarily design standards that describe how a product should be developed and define the particular attributes or characteristics associated with its construction. Alternately, performance standards describe how a product should function and how testing should be used to determine that it meets all affirmed requirements.

The Need for Standards in NG911

It is imperative that the necessary NG911-related standards and technology are determined and available for 911 Authorities and PSAPs to support transitioning to an open, non-proprietary NG911 system. Without the critical standards and technologies in place, service and equipment providers may develop new, vendor-specific solutions. This un-standardized, unplanned approach can and will affect the ability of PSAPs and emergency response entities to effectively share information and be interoperable. Further, without critical processes and protocols (e.g., certification and authentication, routing business rules, and best practices), the benefits of the NG911 system, including routing based on criteria beyond location and connection of service providers beyond common carriers to the 911 system, may not be realized. The appropriate use of standards will ensure the compatibility and interoperability required to realize the full potential of NG911.

Standards Affecting NG911

It is important to identify, understand, and actively monitor those standards that are most likely to have a significant impact on the implementation of NG911. This is consistent with the National Technology Transfer and Advancement Act of 1995[13], which directs government agencies to use "voluntary consensus standards" created by SDOs. Specifically, it instructs federal agencies, such as USDOT, to participate in the standards development process so that these organizations remain aware of USDOT's position on relevant standards. This involvement is expected to influence overall development, thus ensuring that the resulting standard is appropriate for use by federal agencies.

The specific standards identified in this document are limited to those most directly germane to NG911. For example, numerous technical standards are associated with the existing access and originating networks. However, this document undertakes to highlight only those relating to the changes required to support the enhanced capability, such as emergency call support provisioning between the assortment of client devices and Emergency Services IP networks (ESInets). Standards involving network interfaces, including Voice over Packet (VoP), Voice over Internet Protocol (VoIP), or Voice over Digital Subscriber Line (VoDSL), although critical to the end-to-end architecture, are too detailed and non-specific to NG911 for inclusion.

What's New in Standards

Standards and best practices are ever changing to adapt to the current environment. Added in 2016, this new section to the *NG911 Standards Identification and Review* document is included to provide a snapshot of changes that may impact the public safety community. This section is not all inclusive; users are recommended to review any document listed before using it, and should review each document already in use for updates.

3rd Generation Partnership Project (3GPP)

The 3GPP TSG-SA Release 12 included as a priority in this release the use of LTE technology for emergency and security services, with technical specifications for mission-critical application layer functional elements and interfaces with further progress on Mission-Critical Press-to-Talk (MC-PTT) and other features scheduled in Release 13.

American National Standards Institute (ANSI)

The Homeland Security Standards Panel (HSSP) was renamed in 2013 to the Homeland Defense and Security Standardization Collaborative (HDSSC) to include issues related to homeland defense and to include input from a wider range of government agencies and private groups.

---

[13] P.L. 104-113. Available at: http://www.nist.gov/standardsgov/nttaa-act.cfm (last accessed February 24, 2016).

Association of Public-Safety Communication Officials (APCO)

APCO completed several new standards and have several in development:

- New:
    - APCO/NENA ANS 1.107.1.2015: *Standard for the Establishment of a Quality Assurance and Quality Improvement Program for Public Safety Answering Points*
    - APCO ANS 1.110.1-2015: *Multi-Functional Multi-Discipline Computer Aided Dispatch (CAD) Minimum Functional Requirements*
    - APCO/NENA ANS 3.105.1-2015: *Minimum Training Standard for TTY/TDD Use in the Public Safety Communications Center*
    - APCO ANS 3.107.1.2015: *Core Competencies and Minimum Training Requirements for Public Safety Communications Technician*
- In Development:
    - APCO 1.113.1.201x: *Public Safety Communications Call Handling Process*
    - APCO 2.104.1.201x: *Application Integration (for/with) Public Safety Answering Points (PSAPs) and Public Safety Responders*

Alliance for Telecommunications Industry Solutions (ATIS)

Many technical standards developed by ATIS committees such as Emergency Services Interconnection Forum (ESIF), Wireless Technologies and Systems Committee (WTSC), Next Generation Interconnection Interoperability Forum (NGIIF), and Packet Technologies and Systems Committee (PTSC) are related to or being impacted by NG911, such as wireless location accuracy, interim text-to-911 and NG911 service and network interactions. The following standards were recently added and/or updated:

- ATIS-0500027: *Recommendations for Establishing Wide Scale Indoor Location Performance*
- ATIS-0500028: *Analysis of Unwanted User Service Interactions with NG9-1-1 Capabilities*
- ATIS-0700015.v003: *ATIS Standard for Implementation of 3GPP Common IMS Emergency Procedures for IMS Origination and ESInet/Legacy Selective Router Termination*
- ATIS-1000061.2015: *LTE Access Class 14 for National Security and Emergency Preparedness (NS/EP) Communications*
- ATIS-1000065.2015: *Emergency Telecommunications Service (ETS) Evolved Packet Core (EPC) Network Element Requirements*
- ATIS-1000067.2015: *IP NGN Enhanced Calling Name (eCNAM)*
- ATIS-1000679.2015: *Interworking between Session Initiation Protocol (SIP) and ISDN User Part*
- J-STD-110.v002: *Joint ATIS/TIA Native SMS/MMS to 9-1-1 Requirements and Architecture Specification, Release 2*
- J-STD-110.01.v002: *Joint ATIS/TIA Implementation Guideline for J-STD-110, Joint ATIS/TIA Native SMS/MMS to 9-1-1 Requirements and Architecture Specification, Release 2*

Building Industries Consulting Service International (BICSI)

BICSI was added as a new SDO. BICSI is a professional organization that supports the advancement of information and communications technologies. BICSI encompasses the design, integration, and installation of distributions systems and infrastructure, including data centers.

CableLabs

CableLabs was added as a new SDO. CableLabs is a non-profit research and development consortium that includes work on interoperability for cable operators' equipment.

Department of Commerce (DOC)

A new project, National Strategy for Trusted Identities in Cyberspace (NSTIC), was added. NSTIC envisions an online environment—the "Identity Ecosystem"—that improves on the use of passwords and usernames and allows individuals and organizations to better trust one another, with minimized disclosure of personal information.

Department of Homeland Security (DHS)

The National Information Exchange Model (NIEM) was removed from DHS and re-created in its own section.

Department of Justice (DOJ)

NIEM was also removed from DOJ and re-created in its own section.

Department of Transportation (USDOT)

USDOT was updated to reflect the reorganization in the Office of the Assistant Secretary for Research and Technology (OST-R), which is now responsible for all of the program offices, statistics, and research activities previously administered by the RITA.

Emergency Services Workshop (ESW)

ESW was removed as it has been inactive since 2011.

European Telecommunications Standards Institute (ETSI)

ETSI started the Next Generation Protocols (NGP) Industry Specification Group, which is looking at evolving communications and networking protocols to provide the scale, security, mobility, and ease of deployment required for the connected society of the 21st century.

Federal Communications Commission (FCC)

Information was updated on the Network and Reliability Interoperability Council (NRIC) and Communications Security, Reliability, and Interoperability Council (CSRIC), with sections added for ongoing CSRIC work related to NG911.

Information Security Forum (ISF)

ISF was added as a new SDO. ISF is an independent, not-for-profit association developing best practice methodologies, processes, and solutions in cyber, information security, and risk management.

International Academies of Emergency Dispatch (IAED)

Additional certifications were added to IAED.

ISACA

ISACA, previously known as the Information Systems Audit and Control Association, was added as a new SDO. ISACA is an independent, non-profit, global association. ISACA engages in the development, adoption, and use of globally accepted, industry-leading knowledge and practices for information systems. One project is COBIT 5, a business framework for the governance and management of enterprise IT.

National Emergency Number Association (NENA)

NG911 has prompted a review and update of many NENA standards and documents. The following NENA documents have been recently updated or are part of in-progress work:

- NENA 02-014 v1: *GIS Data Collection and Maintenance Standards*
- NENA-STA-006.1: *GIS Data Model for NG9 1-1* (Draft)
- NENA 08-003 v1: *Detailed Functional and Interface Specification for the NENA i3 Solution – Stage 3* (to be renumbered NENA-STA-010; completing second public review first quarter 2016)
- NENA 08-506 v1: *NENA Emergency Services IP Network Design for NG9-1-1 (NID)*
- NENA 71-001 v1: *NENA Standard for NG9-1-1 Additional Data*
- NENA 75-001: *NENA Security for Next-Generation 9-1-1 Standard (NG-SEC)*
- NENA-INF-008.2-2013: *NENA NG9-1-1 Transition Plan Considerations Information Document*
- NENA-INF-014.1-2015: *NENA Information Document for Development of Site/Structure Address Point GIS Data for 9-1-1*
- NENA-INF-TBD: *Non-Mobile Wireless and Broadband Connectivity*
- NENA-INF-TBD: *NENA Classes of Service*
- NENA TBD: *Discrepancy, Performance and Audits for NG9-1-1*
- NENA/APCO-REQ-001.1.1-2016: *NENA/APCO Next Generation 9-1-1 Public Safety Answering Point Requirements*
- NENA-INF-TBD: *Monitoring and Managing NG9-1-1*
- NENA 70-Draft: *Standards for the Provisioning and Maintenance of GIS data to ECRF/LVF*
- NENA-REF-003.1-2015: *NENA Text-to-9-1-1 Public Education*

National Fire Protection Association (NFPA)

NFPA was updated to include the revised NFPA 1221 standard, *Standard for the Installation, Maintenance, and Use of Emergency Services Communications Systems*. The updated version includes changes such as:

- Updated call answering and processing times
- New requirement to have a minimum of two telecommunicators in the PSAP at all times
- Enhanced telecommunicator support during critical incidents
- Two-way radio communications enhancements
- Added text on pathway survivability in conjunction with updated NFPA 72

National Information Exchange Model (NIEM)

NIEM was previously listed under DHS and DOJ as a joint partnership of the two departments. Due to the increased collaboration with the Department of Health and Human Services beginning in 2010, NIEM is now listed as a separate entity.

Network Reliability and Interoperability Council (NRIC)

NRIC documents have been removed to reflect CSRIC replacing NRIC and updating the NRIC documents.

Object Management Group® (OMG®)

OMG® was added as a new SDO. OMG® is an international, open membership, not-for-profit technology standards consortium that develops enterprise integration standards for a wide range of technologies. The OMG work on Cyber Security Protection for Front Line Real-Time Systems and Operational Threat & Risk Model projects may impact NG911.

Standards Coordinating Council (SCC)

SCC was added as a new organization that works with governmental agencies, industry, and SDOs to provide advice and counsel to the standards development stakeholder community on matters related to information sharing and safeguarding standards.

Telcordia

Telcordia was added as a new organization that publishes vendor-neutral technical documentation and roadmaps to implementation of new technologies from the central office perspective. Some of these documents will assist with protecting data centers and equipment in the NG911 environment.

Standards and Best Practices Organizations

This section identifies the work performed and currently underway by professional organizations and SDOs involved with the requirements and specifications pertaining to the implementation of NG911. For each, a summary of the organization includes its purpose (e.g., charter, mission statement), pertinent subgroups within the organization (e.g., committees, working groups), standards involvement, formal activities coordinated with other SDOs, and a statement of the effect of the organization's activities on

NG911 implementation. In each case, the information was reviewed by the respective SDO. Additionally, this information provides perspective on the involvement of 911 within the broader world of emergency response and public safety.

For a more detailed look at individual standards, see Appendix A: Standards and Best Practices.

3rd Generation Partnership Project (3GPP)

Name 3rd Generation Partnership Project (3GPP)

Type International Standards Organization—Industry (Mobile Broadband/Universal Mobile Telecommunications System [UMTS])

Summary 3GPP unites seven telecommunications SDOs (Association of Radio Industries and Businesses [ARIB], Alliance for Telecommunications Industry Solutions [ATIS], China Communications Standards Association [CCSA], European Telecommunications Standards Institute [ETSI], Telecommunications Standards Development Society, India [TSDSI], Telecommunications Technology Association, Korea [TTA], and Telecommunication Technology Committee, Japan [TTC]), known as "Organizational Partners," and provides their members with a stable environment to produce the reports and specifications that define 3GPP technologies.

Purpose The purpose of 3GPP is to prepare, approve, and maintain globally applicable technical specifications and technical reports for:
- An evolved 3rd Generation and beyond Mobile System based on the evolved 3GPP core networks, and the radio access technologies supported by the Partners (i.e., UMTS Terrestrial Radio Access [UTRA] both frequency division duplex [FDD] and time division duplex [TDD] modes), to be transposed by the Organizational Partners into appropriate deliverables (e.g., standards).
- The Global System for Mobile Communications (GSM) including GSM evolved radio access technologies (e.g., General Packet Radio Service [GPRS] and Enhanced Data Rates for GSM Evolution [EDGE]).
- An evolved IP Multimedia Subsystem (IMS) developed in an access independent manner.[14]

Relevant Specification Groups
- TSG CT: The Technical Specification Group (TSG) Core Network and Terminals (CT) is responsible for specifying terminal interfaces (logical and physical), terminal capabilities (e.g., execution environments) and the core network element of 3GPP systems.[15]
- TSG SA: The TSG Service and System Aspects (TSG-SA) is responsible for the overall architecture and service capabilities of systems based on 3GPP specifications and, as such, has a responsibility for cross TSG coordination.[16]

---

[14]3GPP, *Third Generation Partnership Project Agreement.* Available at: http://www.3gpp.org/ftp/Inbox/2008_web_files/3GPP_Scopeando310807.pdf (last accessed February 24, 2016).
[15]3GPP, *CT Plenary Core Networks and Terminals.* Available at: http://www.3gpp.org/CT (last accessed February 24, 2016).
[16] 3GPP, *Service and System Aspects.* Available at: http://www.3gpp.org/-SA- (last accessed February 24, 2016).

Standards

- 3GPP TS 23.167: *3rd Generation Partnership Project; Technical Specification Group Services and System Aspects; IP Multimedia Subsystem (IMS) emergency sessions*
- 3GPP TS 23.228: *3rd Generation Partnership Project; Technical Specification Group Services and System Aspects; IP Multimedia Subsystem (IMS); Stage 2*
- 3GPP TS 23.517: *3rd Generation Partnership Project; Technical Specification Group Services and System Aspects; IP Multimedia Subsystem (IMS); Functional Architecture*
- 3GPP TS 24.229: *3rd Generation Partnership Project; Technical Specification Group Core Network and Terminals; IP multimedia call control protocol based on Session Initiation Protocol (SIP) and Session Description Protocol (SDP); Stage 3*
- 3GPP TS 29.010: *3rd Generation Partnership Project; Technical Specification Group Core Network and Terminals; Information element mapping between Mobile Station - Base Station System (MS - BSS) and Base Station System - Mobile-services Switching Centre (BSS - MSC); Signaling procedures and the Mobile Application Part (MAP)*
- 3GPP TSG SA Release 12*: 3rd Generation Partnership Project; Exploits new business opportunities such as Public Safety and Critical Communications, explores Wi-Fi integration, and system capacity and stability*
- 3GPP TSG SA Release 13: *3rd Generation Partnership Project; 3GPP is considering radio technologies to meet the requirements of very low power consumption and a reduced burden on the network from the growing number of terminals on the IoT* [17]
- 3GPP TSG SA Release 14: *3rd Generation Partnership Project; Focusing at this early stage on Mission Critical enhancements, long-term evolution (LTE) support for V2x services, eLAA, 4-band Carrier Aggregation, inter-band Carrier Aggregation and more*[18]

Coordinated Activities

- Open Mobile Alliance (OMA): Based on the "OMA-3GPP Standardization Collaboration," OMA and 3GPP will work to update on a regular basis the list of dependencies between each organization's specifications and work in progress.[19]

---

[17] 3GPP Release 13 Available at: http://www.3gpp.org/release-13 (last accessed February 24, 2016).
[18] 3GPP Release 14 Available at: http://www.3gpp.org/release-14 (last accessed February 24, 2016).
[19] Open Mobile Alliance, *3GPP Dependencies.* Available at: http://www.openmobilealliance.org/Technical/3gpp.aspx (last accessed February 26, 2016).

Effects on NG911

- Develops standards that enable text and multimedia transmission from the caller to the NG911 system (transport of data).
- Develops standards adhered to by originating service providers' (OSP) network and applications services for emergency calling.
- Supports location requirements and standards.

Website   http://www.3gpp.org/

3rd Generation Partnership Project 2 (3GPP2)

Name   3rd Generation Partnership Project 2 (3GPP2)

Type   International Standards Organization—Industry (Mobile Broadband/UMTS)

Summary   3GPP2 is a collaboration among groups of telecommunications associations to develop a globally applicable 3G mobile telephone system specification within the scope of the IMT-2000 project of the International Telecommunication Union (ITU). 3GPP2 specifications are based on the Code Division Multiple Access 2000 (CDMA2000) 3G mobile technology standards. 3GPP2 can be characterized as a collaborative 3G telecommunications specifications-setting project:

- Comprising North American and Asian interests developing global specifications for ANSI/Telecommunications Industry Association (TIA)/Electronics Industry Alliance (EIA)-41 Cellular Radio Telecommunication Intersystem Operations network evolution to 3G.
- Developing global specifications for the radio transmission technologies supported by ANSI/TIA/EIA-41.

3GPP2 was born out of the ITU's IMT-2000 initiative covering high-speed, broadband, and IP-based mobile systems featuring network-to-network interconnection, feature/service transparency, global roaming, and seamless services independent of location. IMT-2000 is intended to bring high-quality mobile multimedia telecommunications to a worldwide mass market by achieving the goals of increasing the speed and ease of wireless communications, responding to the problems faced by the increased demand to pass data via telecommunications, and providing "anytime, anywhere" services.[20]

Relevant Specification Groups
- TSG-AC: The Access Network & Air Interfaces Technical Specification Group (TSG-AC) is responsible for the radio access part, including its internal structure; the specification of interfaces and transports between the radio access network and core network, as well as within the access network; the interworking between 3GPP2 technologies and with other radio access technologies.[21]

---

[20] 3GPP2, *About 3GPP2: What is 3GPP2?* Available at: http://www.3gpp2.org/Public_html/Misc/AboutHome.cfm (last accessed February 24, 2016).

[21] 3GPP2, *TSG-AC: Access Network & Air Interfaces*. Available at: http://www.3gpp2.org/public_html/AC/index.cfm (last accessed March 1, 2016).

Relevant Specification Groups (continued)	• <u>TSG-SX</u>: The System Aspects & Core Network Technical Specification Group (TSG-SX) is responsible for the development of service capability requirements for 3GPP2 systems; high level architectural issues, as required, to coordinate service development across the various TSGs; and specification of the core network part of 3GPP2 systems.[22]
Standards	• 3GPP2 S.R0006-529-A: *Wireless Features Description: Emergency Services* • 3GPP2 X.S0049-0: *All-IP Network Emergency Call Support* • 3GPP2 X.S0057-A: *E-UTRAN - eHRPD Connectivity and Interworking: Core Network Aspects* • 3GPP2 X.S0060-0: *HRPD Support for Emergency Services*
Coordinated Activities	• OMA: Based on the OMA-3GPP2 Standardization Collaboration, OMA and 3GPP2 will work to update on a regular basis the list of dependencies between each organization's specifications and work in progress.[23] • TIA: 3GPP2 is a collaborative effort among five officially recognized SDOs—ARIB, CCSA, TTA, TTC, and TIA.
Effects on NG911	• Develops standards that enable text and multimedia transmission from the caller to the NG911 system (transport of data). • Supports location requirements and standards.
Cross-Reference to Published Standards	• 3GPP2 specifications and reports are converted into standards by each of the Project's Organizational Partners. http://www.3gpp2.org/Public_html/specs/index.cfm
Website	http://www.3gpp2.org/

---

[22] 3GPP2, *TSG-X Core Networks*. Available at: http://www.3gpp2.org/public_html/SX/index.cfm (last accessed March 1, 2016).

[23] Open Mobile Alliance, *3GPP2 Dependencies*. Available at: http://www.openmobilealliance.org/Technical/3gpp2.aspx (last accessed February 26, 2016).

American National Standards Institute (ANSI)

Name

American National Standards Institute (ANSI)

Type

National Standards Organization

Summary

ANSI is a private, not-for-profit organization that oversees development of voluntary consensus standards in the U.S. Activities include accrediting programs, assessing conformance, and approving standards developed by organizations such as ATIS and the Association of Public-Safety Communications Officials, International (APCO). ANSI, itself, does not set standards, but approves and accredits other SDOs. Membership is composed of government agencies, academic and international bodies, and individuals. ANSI is the official U.S. representative to the ISO and, via the U.S. National Committee, the IEC.

Mission

ANSI's mission is to enhance both the global competitiveness of U.S. business and the U.S. quality of life by promoting and facilitating voluntary consensus standards and conformity assessment systems, and safeguarding their integrity.[24]

Relevant Standards Panel

- Homeland Defense and Security Standardizations Collaborative (HDSSC): ANSI-HDSSC has as its mission to identify existing consensus standards, or, if none exist, assist government agencies and those sectors requesting assistance to accelerate development and adoption of consensus standards critical to homeland security and homeland defense. Originally established by ANSI in February 2003 as the HSSP, the HDSSC was renamed in 2013 to reflect a revision of the group's charter expanding its scope to include issues related to homeland defense and to include input from a wider range of government agencies and private groups. The HSSP was originally established to bolster the development of voluntary standards related to homeland security and emergency preparedness. ANSI-HDSSC promotes a positive, cooperative partnership between the public and private sectors to meet the needs of the nation in this critical area.[25]

---

[24] American National Standards Institute, *About ANSI Overview.* Available at: http://www.ansi.org/about_ansi/overview/overview.aspx (last accessed February 24, 2016).
[25] ANSI Standards Activities, *Homeland Defense and Security Standardization Collaborative.* Available at: http://www.ansi.org/standards_activities/standards_boards_panels/hssp/overview.aspx?menuid=3 (last accessed February 24, 2016).

Coordinated Activities	• National Institute of Standards and Technology (NIST): A Memorandum of Understanding (MOU) exists between NIST and ANSI that agrees on the need for a unified national approach to develop the best possible national and international standards.[26] • ISO: ANSI is the sole U.S. representative and dues-paying member of the ISO. As a founding member of the ISO, ANSI plays a strong leadership role in its governing body.[27]
Effects on NG911	• Validates the standards development process for SDOs that produce standards affecting NG911.
Website	http://www.ansi.org/

---

[26] National Institute of Standards of Technology Standards Coordination and Conformity Group, *Memorandum of Understanding between the American National Standards Institute (ANSI) and the National Institute of Standards and Technology (NIST)*. Available at: http://gsi.nist.gov/global/docs/ANSINISTMOU2000.pdf (last accessed February 24, 2016).

[27] American National Standards Institute, *ANSI Accredited of U.S. Technical Advisory Groups (TAGs) to ISO*. Available at: http://www.ansi.org/standards_activities/iso_programs/tag_iso.aspx (last accessed February 24, 2016).

Association of Public-Safety Communications Officials (APCO)

Name	Association of Public-Safety Communications Officials-International (APCO)
Type	National Standards Organization (ANSI-accredited)
Summary	APCO is the world's oldest and largest organization dedicated to public safety communications and is an ANSI-accredited SDO committed to ensuring public safety communications personnel have a role in the development of standards that affect the industry. APCO's standards development activities have a broad scope, ranging from actual development of standards to representation of public safety communications organizations in other standards development areas.[28] APCO International develops standards and disseminates information about critical issues such as wireless 9-1-1, staffing and retention, and the impact of emerging technologies. APCO participates in numerous committees, partnerships, and government initiatives. APCO supports agencies around the country grappling with the industry's toughest issues by delivering a variety of resources and engaging in the latest research to find common solutions.[29]
Mission	APCO is an international leader committed to providing complete public safety communications expertise, professional development, technical assistance, advocacy and outreach to benefit our members and the public.
Relevant Committees	• 9-1-1 Emerging Technologies: The 9-1-1 Emerging Technologies Committee identifies issues and makes recommendations to the standards development for data delivery in an all IP environment. This committee provides subject-matter experts to the International Committee related to U.S. 9-1-1 issues, has established at least two strategic alliances related to the mission of APCO, provides leadership opportunities for committee members by establishing work groups within the 9-1-1 Emerging Technologies Committee, and has established a 9-1-1 public policy work group to identify key areas of public policy that APCO should influence or advocate for related 9-1-1 operations.[30]

---

[28] APCO, *About APCO.* Available at: http://apcointl.org/about-apco.html and https://www.apcointl.org/standards.html (last accessed February 24, 2016).
[29] APCO, *911 Resources.* Available at: http://apcointl.org/resources.html (last accessed February 24, 2016).
[30] APCO, *9-1-1 Emerging Technologies Committee.* Available at: https://apconetforum.org/eweb/DynamicPage.aspx?Webcode=APCOCommDescript&APCOcmt_key=11e96d6f-46f8-4044-be27-a7aa8233b72f (last accessed February 26, 2015).

Relevant Projects

- Project 25: A joint effort of APCO and the National Association of State Telecommunications Directors, Project 25 concerns the development of standards for digital telecommunications technology, including an objective to determine consensus standards for digital radio equipment embracing elements of interoperability, spectrum efficiency, and cost economies.[31]
- Project 36: This project was developed to research and develop universal standards for computer aided dispatch (CAD) and CAD-to-CAD exchanges. The goal was to develop effective processes for the exchange of data between third-party call centers such as alarm companies and PSAPs.[32]
- Project 42 (Global Operating Picture): The goal of Project 42 is to identify those areas where standards are needed to achieve system interoperability and create a common operating picture at all levels, horizontal and vertical.[33]
- Project 43 (Broadband Implications for the PSAP): The goal of Project 43 is to help telecommunicators, PSAPs, 9-1-1 authorities, emergency operations centers, and others prepare for evolving broadband communications technologies that will impact PSAP operations and support emergency responders.[34]

Standards

- APCO/NENA ANS 1.107.1-2015: *Standard for the Establishment of a Quality Assurance and Quality Improvement Program for Public Safety Answering Points*
- APCO ANS 1.116.1-2015: *Public Safety Communications Common Status Codes for Data Exchange*
- APCO ANS 1.112.1-2014: *Best Practices for The Use of Social Media in Public Safety Communications*
- APCO ANS 1.110.1-2015: *Multi-Functional Multi-Discipline Computer Aided Dispatch (CAD) Minimum Functional Requirements*
- APCO/NPSTC ANS 1.104.1-2010: *Standard Channel Nomenclature for the Public Safety Interoperability Channels*
- APCO ANS 1.101.3-2015: *Standard for Public Safety Telecommunicators When Responding to Calls of Missing, Abducted and Sexually Exploited Children*
- APCO/NENA ANS 1.105.2-2015: *Standard for Telecommunicator Emergency Response Taskforce (TERT) Deployment*
- APCO ANS 3.103.2-2013: *Wireless 9-1-1 Deployment and Management Effective Practices Guide*
- APCO ANS 1.111.1-2013: *Public Safety Communications Common Disposition Codes for Data Exchange*

---

[31] APCO, *911 Resources APCO Projects.* Available at: http://apcointl.org/about-apco/apco-projects.html (last accessed February 24, 2016).

[32] APCO, *911 Resources APCO Projects.* Available at: http://apcointl.org/about-apco/apco-projects.html (last accessed February 24, 2016).

[33] APCO, *911 Resources APCO Projects.* Available at: http://apcointl.org/about-apco/apco-projects.html (last accessed February 24, 2016).

[34] APCO, *APCO Project 43.* Available at: http://apconetforum.org/eweb/DynamicPage.aspx?WebCode=APCOProject43 (last accessed March 16, 2016).

Standards (continued)

- APCO/CSAA ANS 2.101.2-2014: *Alarm Monitoring Company to Public Safety Answering Point (PSAP) Computer-Aided Dispatch (CAD) Automated Secure Alarm Protocol (ASAP)*
- APCO ANS 2.103.1-2012: *Public Safety Communications Common Incident Types For Data Exchange*
- APCO ANS 3.101.2-2013: *Core Competencies and Minimum Training Standards for Public Safety Communications Training Officer (CTO)*
- APCO ANS 3.108.1.2014: *Core Competencies and Minimum Training Standards for Public Safety Communications Instructor*
- APCO ANS 3.106.1-2013: *Core Competencies and Minimum Training Standards for Public Safety Communications Quality Assurance Evaluator (QAE)*
- APCO ANS 3.102.1-2012: *Core Competencies and Minimum Training Standards for Public Safety Communications Supervisor*
- APCO ANS 3.109.2.2014: *Core Competencies and Minimum Training Standards for Public Safety Communications Manager/Director*
- APCO ANS 3.104.1-2012 : *Core Competencies and Minimum Training Standards for Public Safety Communications Training Coordinator*
- APCO ANS 3.103.2.2015: *Minimum Training Standards for Public Safety Telecommunicators*
- APCO ANS 3.107.1.2015: *Core Competencies and Minimum Training Requirements for Public Safety Communications Technician*
- APCO/NENA ANS 3.105.1-2015: *Minimum Training Standard for TTY/TDD Use in the Public Safety Communications Center*
- APCO/NENA ANS 1.102.2-2010: *Public Safety Answering Point (PSAP) Service Capability Criteria Rating Scale* (In Revision)
- APCO 1.108.1-201x: *Minimum Operational Standards for the Use of TTY/TDD devices in the Public Safety Communications Center* (In Progress)
- APCO 1.113.1-201x: *Public Safety Communications Call Handling Process* (In Progress)
- APCO 1.114.1-201x: *Vehicle Telematics Best Practices* (In Progress)
- APCO 1.115.1-201x: *Core Competencies, Operational Factors, and Training for Next Generation Technologies in Public Safety Communications* (In Progress)
- APCO 2.102.1.201x: *Advanced Automatic Collision Notification (AACN) Data Set* (In Progress)
- APCO 2.104.1.201x: *Application Integration (for/with) Public Safety Answering Points (PSAPs) and Public Safety Responders (In Progress)*
- APCO/NENA 2.105.1201x *NG9-1-1 Emergency Incident Data Document (EIDD)* (In Progress)

Coordinated Activities

- ANSI: As an ANSI-accredited Standards Developer (ASD), APCO is dedicated to ensuring public safety communications personnel have a role in the development of standards that affect communications professionals.[35]

Websites

http://www.apcointl.org/
http://www.apcostandards.org/

---

[35] APCO, *911 Resources.* Available at: http://apcointl.org/standards.html (last accessed February 26, 2015).

Alliance for Telecommunications Industry Solutions (ATIS)

Name   Alliance for Telecommunications Industry Solutions (ATIS)

Type   Standards Setting Organization—Industry (Telecommunications)

Summary   ATIS is a standards organization that develops technical and operational standards for the telecommunications industry. Member companies include telecommunications service providers, equipment manufacturers, public sector entities, and others. ATIS is accredited by ANSI; is a member organization of other standards organizations, including the Radiocommunication Sector (ITU-R) and Standardization Sector (ITU-T) of the ITU; and is an Organizational Partner of 3GPP.

The priorities that ATIS is currently addressing include the following:
- Advancing the 5G network, with a focus on North American requirements contributing to a global 5G standard
- Bringing a comprehensive problem-solving approach to the all-IP network transition and ensuring it proceeds at the desired pace of the industry
- Creating solutions and an overall industry framework for addressing cybersecurity threats
- Developing open source solutions in the context of an interoperable standards environment
- Creating the next-generation emergency communications advances that the market demands[36]

Relevant Committees/ Subcommittees
- Emergency Services Interconnection Forum (ESIF): ESIF, composed of wireless and wireline network service providers, manufacturers, public sector entities, and providers of support services, facilitates identification and resolution of technical issues related to the interconnection of telephony and emergency services networks.[37]
  - Next Generation Emergency Services (NGES) Subcommittee: The NGES Subcommittee coordinates emergency services needs and issues with and among SDOs and industry forums/committees, and within and outside ATIS; and develops emergency services (e.g., Enhanced 9-1-1 [E9-1-1]) standards and other documentation related to advanced (i.e., next generation) emergency services architectures, functions, and interfaces for communications networks.[38]

---

[36] ATIS, *About ATIS.* Available at: http://www.atis.org/about/ (last accessed February 26, 2015).
[37] ATIS, *Emergency Services Interconnection Forum.* Available at: http://www.atis.org/01_committ_forums/ESIF/index.asp (last accessed February 25, 2016).
[38] ATIS, *Next Generation Emergency Services (NGES). Subcommittee.* Available at: http://www.atis.org/esif/nges.asp (last accessed February 25, 2016).

Relevant Committees/ Subcommittees (continued)

- o Emergency Services & Methodologies (ESM) Subcommittee: The mission of the ESIF ESM Subcommittee is to provide a set of minimum, practical requirements that will ensure consistent, valid, verifiable, and reproducible location data in a variety of access environments based on sound engineering and statistical practice.[39]
- Next Generation Interconnection Interoperability Forum (NGIIF): NGIIF addresses next generation network interconnection and interoperability issues associated with emerging technologies. It develops operating procedures that involve the network aspects of architecture, disaster preparedness, installation, maintenance, management, reliability, routing, security, and testing between network operators, with a current focus on call completion.[40]
- Packet Technologies and Systems Committee (PTSC): PTSC develops and recommends standards and technical reports related to packet services and packet service architectures, in addition to related subjects under consideration in other North American and international standards bodies.[41]
- Wireless Technologies and Systems Committee (WTSC): WTSC develops and recommends standards and technical reports related to wireless and/or mobile services and systems, including service descriptions and wireless technologies. WTSC also develops and recommends positions on related subjects under consideration in other North American, regional, and international standards bodies.[42]
- Telecom Management and Operations Committee (TMOC): The TMOC develops operations, administration, maintenance and provisioning standards, and other documentation related to Operations Support System (OSS) and Network Element (NE) functions and interfaces for communications networks - with an emphasis on standards development related to U.S. communication networks in coordination with the development of international standards.[43]

---

[39] ATIS, *Emergency Services & Methodologies (ESM) Subcommittee.* Available at: http://www.atis.org/ESIF/esifsubcommitteeg.asp (last accessed February 25, 2016)
[40] ATIS, *NGIIF: Next Generation Interconnection Interoperability Forum.* Available at: http://www.atis.org/01_committ_forums/NGIIF/index.asp (last accessed February 25, 2016).
[41] ATIS, *Packet Technologies and Systems Committee (PTSC).* Available at: http://www.atis.org/01_committ_forums/PTSC/index.asp (last accessed February 25, 2016).
[42] ATIS, *Wireless Technologies and Systems Committee (WTSC).* Available at: http://www.atis.org/01_committ_forums/WTSC/index.asp (last accessed February 25, 2016).
[43] ATIS, *Telecom Management and Operations Committee (TMOC).* Available at: http://www.atis.org/01_committ_forums/TMOC/index.asp (last accessed February 25, 2016).

Standards

- ATIS-0100022.2008(R2013): *Priority Classification Levels for Next Generation Networks*
- ATIS-0300104: *Next Generation Interconnection Interoperability Forum (NGIIF) NGN Reference Document - NGN Basics, Emergency Services, NGN Testing, and Network Survivability*
- ATIS-0500001: *High Level Requirements for Accuracy Testing Methodologies*
- ATIS-0500002.2008(R2013): *Emergency Services Messaging Interface (ESMI)*
- ATIS-0500003: *Routing Number Authority (RNA) for pseudo Automatic Number Identification Codes (pANIs) Used for Routing Emergency Calls: pANI Assignment Guidelines and Procedures*
- ATIS-0500004: *Recommendation for the Use of Confidence and Uncertainty for Wireless Phase II*
- ATIS-0500005: *Standard Wireless Text Message Case Matrix*
- ATIS-0500006.2008(R2013): *Emergency Information Services Interfaces (EISI) ALI Service*
- ATIS-0500007.2008: *Emergency Information Services Interface (EISI) Implemented with Web Services*
- ATIS-0500008: *Emergency Services Network Interfaces (ESNI) Framework*
- ATIS-0500009: *High Level Requirements for End-to-End Functional Testing*
- ATIS-0500013: *Approaches to Wireless E9-1-1 Indoor Location Performance Testing*
- ATIS-0500015.2010: *Flexible LDF-AMF (Location Determination Function – Access Measurement Function) Protocol (FLAP) Specification*
- ATIS-0500017: Technical Report: *Considerations for an Emergency Services Next Generation Network (ES-NGN)*
- ATIS-0500018: *P-ANI Allocation Tables for ESQKs, ESRKs, and ESRDs*
- ATIS-0500019.2010: *Request for Assistance Interface (RFAI) Specification*
- ATIS-0500021: *Supplemental Location Data*
- ATIS-0500022: *Test Plan Input for a Location Technology Test Bed*
- ATIS-0500023: *Applying Common IMS to NG9-1-1 Networks*
- ATIS-0500024: Technical Report: *Comparison of SIP Profiles*
- ATIS-0500025: *Class of Service Support for Semi-Static Wireless*
- ATIS-0500026: *Operational Impacts on Public Safety of ATIS-0700015, Implementation of 3GPP Common IMS Emergency Procedures for IMS Origination and ESInet/Legacy Selective Router Termination*
- ATIS-0500027: *Recommendations for Establishing Wide Scale Indoor Location Performance*
- ATIS-0500028: Technical Report: *Analysis of Unwanted User Service Interactions with NG9-1-1 Capabilities*
- ATIS-0700015.v003: *ATIS Standard for Implementation of 3GPP Common IMS Emergency Procedures for IMS Origination and ESInet/Legacy Selective Router Termination*

Standards (continued)

- ATIS-1000010.2006(R2011): *Support of Emergency Telecommunications Service ETS in IP Network*
- *ATIS-1000012.2006: Signaling System No. 7 (SS7) – SS7 Network and NNI Interconnection Security Requirements and Guidelines*
- *ATIS-1000019: Network to Network Interface (NNI) Standard for Signaling and Control Security for Evolving VoP Multimedia Networks*
- ATIS-1000023.2013: *ETS Network Element Requirements for A NGN IMS Based Deployments*
- ATIS-1000026.2008(R2013): *Session Border Controller Functions and Requirements*
- ATIS-1000029.2008: *Security Requirements for NGN*
- ATIS-1000034.2010(R2015): *Next Generation Network (NGN): Security Mechanisms and Procedures*
- *ATIS-1000038: Technical Parameters for IP Network to Network Interconnection Release 1.0*
- *ATIS-1000040: Protocol Suite Profile for IP Network to Network Interconnection Release 1.0*
- *ATIS-1000041: Test Suites for IP Network to Network Interconnection Release 1.0*
- ATIS-1000049: *End-to-End NGN GETS Call Flows*
- ATIS-1000055.2013: *Emergency Telecommunications Service (ETS): Core Network Security Requirements*
- ATIS-1000060.2014: *Emergency Telecommunications Service (ETS): Long Term Evolution (LTE) Access Network Security Requirements for National Security/Emergency Preparedness (NS/EP) Next Generation Network (NGN) Priority Services*
- ATIS-1000061.2015: *LTE Access Class 14 for National Security and Emergency Preparedness (NS/EP) Communications*
- ATIS-1000065.2015: *Emergency Telecommunications Service (ETS) Evolved Packet Core (EPC) Network Element Requirements*
- ATIS-1000067.2015: *IP NGN Enhanced Calling Name (eCNAM)*
- ATIS-1000679.2015: *Interworking between Session Initiation Protocol (SIP) and ISDN User Part*
- ESIF Issue 81: *Applying Common IMS to NG9-1-1 Networks (Stage 2 & 3 Specification)* (In Development)
- ESIF Issue 82: *IMS-based Next Generation Emergency Services Network Interconnection*
- ESIF Issue 86: *Technical Report to describe ATIS-0700015 for Public Safety*
- ESIF Issue 87: *Vertical Axis Measurement Test Methodology*
- ANSI/J-STD-036-C: *Enhanced Wireless 9-1-1 Phase 2*
- ANSI/J-STD-036-C-1: *Addendum to J-STD-036-C- Enhanced Wireless 9-1-1 Phase 2*
- J-STD-110.v002: *Joint ATIS/TIA Native SMS/MMS to 9-1-1 Requirements & Architecture Specification*

Standards (continued)

- J-STD-110.01.v002: *Joint ATIS/TIA Implementation Guideline for J-STD-110, Joint ATIS/TIA Native SMS/MMS to 9-1-1 Requirements and Architecture Specification, Release 2*
- ESIF-E911-Phase2ReadinessPackage: *Wireless E9-1-1 Phase II Readiness Package*
- NGIIF Issue 27: *Documentation of Operational Procedures for Next Generation Networks Interconnection*
- NGIIF Issue 31: *Develop New Text Related to Methodologies That Support TDM/IP Caller ID Services, Call Spoofing, Etc.*
- PTSC Issue 28: *US Standard For IP-IP Network Interconnection - Roadmap Standard*
- PTSC Issue 66: *NGN Architecture Phase 2*
- PTSC Issue 81: *ETS Wireline Access Requirements*
- PTSC Issue 82: *ETS Phase 2 Network Element Requirements*
- PTSC Issue 93: *NGN Security Planning & Operations Guidelines*
- PTSC Issue 98: *ETS Roadmap*
- PTSC Issue 100: *Supplement to ATIS-1000010*
- PTSC Issue 119: *Dynamic Priority for Next Generation Secure Communications*
- WTSC Issue 32: *Support of Public Safety Requirements in LTE Networks*
- WTSC Issue 34: *Automating Location Acquisition for Non-Operator-Managed Over-the-Top VoIP Emergency Services Calls*
- WTSC Issue 39: *Public Safety Mission Critical Push to Talk (PTT) Voice Interoperation between Land Mobile Radio (LMR) and Long Term Evolution (LTE) Systems*
- WTSC Issue 41: *Commercial Mobile Alerts Service (CMAS) International Roaming*
- WTSC Issue 51: *Location Accuracy Improvements for Emergency Calls*

Coordinated Activities

- 3GPP, ETSI, ITU, and NENA: The NGES Subcommittee emphasizes standards development as it relates to North American communications networks, in coordination with the development of standards activities, including relevant ATIS committees (e.g., PTSC), ITU, 3GPP, ETSI, and NENA. [44]
- ANSI: ATIS is an ANSI-accredited SDO.[45]
- TIA: An MOU exists between ATIS and TIA to jointly sponsor and work cooperatively in the development of joint standards documents that are of mutual interest.[46]

---

[44] ATIS, *NGES: Next Generation Emergency Services Subcommittee.* Available at: http://www.atis.org/esif/nges.asp (last accessed February 25, 2016).

[45] ANSI, *ANSI Accredited Standards Developers.* Available at: http://publicaa.ansi.org/sites/apdl/Lists/American%20National%20Standards/AllItems.aspx (last accessed February 25, 2016).

[46] ATIS, *General Principles in Sponsorship of Joint Standards Activities Between the Alliance for Telecommunications Industry Solutions (ATIS) and the Telecommunications Industry Association (TIA).* Available at: http://www.atis.org/legal/Docs/MOU/TIA.pdf (last accessed February 25, 2016).

Effects on NG911

- Develops standards adhered to by OSP's network and applications services for emergency calling.
- Supports location requirements and standards.

Websites

http://www.atis.org/

Broadband Forum (BBF)

Name	Broadband Forum (BBF)
Type	Industry (Broadband)
Summary	BBF is the central organization driving broadband wireline solutions and empowering converged packet networks worldwide to better meet the needs of vendors, service providers, and their customers.
Mission	BBF's mission is to develop multi-service broadband packet networking specifications addressing interoperability, architecture, and management. BBF's work enables home, business, and converged broadband services, encompassing customer, access, and backbone networks. [47]
Relevant Working Groups	<ul><li>End to End Architecture: This group's mission is to oversee and coordinate all access architecture and transport-related technical work within the Forum. Scope includes access architecture encompassing interface definitions and nodal functional requirements—from residential gateway (RG) through access node, aggregation network, and broadband network gateway (BNG) to peering interfaces with network and application service providers. The focus is end-to-end service delivery across this domain encompassing equipment requirements to support capabilities such as quality of service (QoS) and multicast functionality. Working group interests also encompass policy and control of the key network elements and protocol interworking requirements. All broadband wireline access technologies are within the scope of this access architecture work (e.g., Digital Subscriber Line [DSL], Gigabit Passive Optical Network [GPON], and point-to-point fiber). Wireless broadband access technologies are addressed via liaison with the appropriate standards body (e.g., WiMAX Forum, 3GPP, etc.). Consideration is also given to the relative energy efficiency aspects of access architectures.[48]</li><li>BroadbandHome™: This group's mission is to provide the broadband industry with technical specifications that define broadband enabled customer devices and ease the deployment and management of broadband services. [49]</li></ul>

---

[47] Broadband Forum, *Technical Working Groups.* Available at: https://www.broadband-forum.org/about/mission.php (last accessed February 25, 2016).

[48] Broadband Forum, *Technical Working Groups.* Available at: http://www.broadband-forum.org/technical/technicalworkinggroups.php (last accessed February 25, 2016).

[49] Broadband Forum, *Technical Working Groups.* Available at: http://www.broadband-forum.org/technical/technicalworkinggroups.php (last accessed February 25, 2016).

Coordinated Activities	• WiMAX Forum, 3GPP: BBF works alongside mobile-related partners to ensure all their work is aligned.[50]
Website	http://www.broadband-forum.org/

---

[50] Broadband Forum, *Liaising and Collaborating With The Broadband Forum.* Available at: https://www.broadband-forum.org/about/liaisonprogram.php (last accessed February 25, 2016).

Building Industries Consulting Service International (BICSI)

Name	Building Industries Consulting Service International (BICSI)
Type	International Trade Association (Infrastructure Systems)
Summary	BICSI is a professional association supporting the advancement of the information and communications technology (ICT) community. ICT covers the spectrum of voice, data, electronic safety and security, project management, and audio and video technologies. It encompasses the design, integration, and installation of pathways, spaces, optical fiber- and copper-based distribution systems, wireless-based systems, and infrastructure that support the transportation of information and associated signaling between and among communications and information gathering devices.
	BICSI provides information, education, and knowledge assessment for individuals and companies in the ICT industry. BICSI serves nearly 23,000 ICT professionals, including designers, installers, and technicians. These individuals provide the fundamental infrastructure for telecommunications, audio-video, life safety, and automation systems. Through courses, conferences, publications and professional registration programs, BICSI staff and volunteers assist ICT professionals in delivering critical products and services, and offer opportunities for continual improvement and enhanced professional stature.
	BICSI membership spans nearly 100 countries.[51]
Relevant Technical Subcommittees	BICSI Standards Program Technical Subcommittees: The majority of work within the BICSI International Standards Program is performed by its technical subcommittees. Each subcommittee is comprised of technical experts, and is the primary consensus body of the program. Thus, technical subcommittees and their leaders have a great responsibility, as the technical actions and decisions of a technical subcommittee are the actions and decisions for the entire BICSI Standards Program in that subcommittee's field of expertise.

---

[51]BICSI, Available at: http://www.bicsi.org (last accessed February 25, 2016).

Relevant Technical Subcommittees (continued)	The BICSI Standards Program currently has the following subcommittees[52]: • BIM Best Practices • Bonding and Grounding • Codes • Data Center • Educational Facilities • Electronic Safety and Security (ESS) • Healthcare • Intelligent Buildings • International Cabling Installation • North American Cabling Installation • OSP Cabling Installation • Residential • Wireless Systems
Standards	• ANSI/BICSI 002-2014: *Data Center Design and Implementation Best Practices* • ANSI/BICSI 003-2014: *Building Information Modeling (BIM) Practices for Information Technology Systems* • ANSI/BICSI 005-2013: *Electronic Safety and Security (ESS) System Design and Implementation Best Practices* • ANSI/BICSI 006-2015: *Distributed Antenna System (DAS) Design and Implementation Best Practices* • ANSI/NECA/BICSI 568-2006: *Standard for Installing Commercial Building Telecommunications Cabling* • ANSI/NECA/BICSI 607-2011: *Standard for Telecommunications Bonding and Grounding Planning and Installation Methods for Commercial Buildings* • BICSI *Telecommunications Distribution Methods Manual*, 13th Edition • BICSI *Information Technology Systems Installation Methods Manual*, 6th Edition • BICSI *Outside Plant Design Reference Manual*, 5th Edition
Effects on NG911	• Develops standards adhered to by service providers, network engineers and facility managers for emergency calling facilities.
Website	http://www.bicsi.org

---

[52] BISCI, *Standards Program Technical Subcommittees*. Available at: https://www.bicsi.org/double.aspx?l=3316 (last accessed February 25, 2016).

CableLabs

Name	CableLabs
Type	Standards Setting Organization—Industry (Cable)
Summary	CableLabs is a non-profit research and development consortium dedicated to creating innovative ideas that significantly impact our cable operator members' business. CableLabs also serves to define interoperable solutions among our members and their technology vendors in order to drive scale, reduce costs, and create competition in the supply chain. CableLabs membership is comprised of the major cable operators worldwide.[53]
Mission	CableLabs' mission is to deliver innovations that enable our members to be the provider of choice in their markets.
Standards	• Invention Disclosure 60399: *Phased Array Scanner for Fire and Police Applications* • Invention Disclosure 60620: *Terrestrial Wi-Fi based Auto Security and Car Safety Service*
Coordinated Activities	• Working in cooperation with cable operators and cable equipment manufacturers, CableLabs has developed various specifications to facilitate the manufacture of interoperable cable devices, including telephony. "CableLabs Certified®" or "CableLabs Qualified" means that the device has passed a series of tests for compliance with the applicable specification, and has thus demonstrated interoperable functionality with other CableLabs-certified devices. Interoperable devices based on common specifications facilitate consumer choice, wide spread deployment of new technologies, and lower costs to both cable operators and consumers.[54] • 3GPP – 5G networks
Website	http://www.cablelabs.com

---

[53] CableLabs, *About CableLabs*. Available at: http://www.cablelabs.com/about-cablelabs/ (last accessed February 25, 2016).

[54] CableLabs, *CableLabs Certification Program*. Available at: http://www.cablelabs.com/specs/certification/ (last accessed February 25, 2016).

Commission on Accreditation for Law Enforcement Agencies (CALEA)

Name   Commission on Accreditation for Law Enforcement Agencies (CALEA®)

Type   Professional Organization

Summary   CALEA® was created as a credentialing authority through the joint efforts of law enforcement's major executive associations—International Association of Chiefs of Police (IACP), National Organization of Black Law Enforcement Executives (NOBLE), National Sheriffs' Association (NSA), and the Police Executive Research Forum (PERF).

The purpose of CALEA®'s Accreditation Program is to improve the delivery of public safety services, primarily by maintaining a body of standards, developed by public safety practitioners, that covers a wide range of up-to-date public safety initiatives; establishing and administering an accreditation process; and recognizing professional excellence.[55]

APCO partnered with CALEA® to set the *Standards for Public Safety Communications Agencies ©*.

Relevant Committees   • Standards Review and Interpretation Committee (SRIC)

Standards   • CALEA® *Standards for Law Enforcement Agencies©*
     • CALEA® *Standards for Public Safety Communications Agencies©*

Website   http://www.calea.org/

---

[55]CALEA, *About Us,* Available at: http://www.calea.org/content/commission (last accessed February 25, 2016).

Department of Commerce (DOC)

Name	Department of Commerce (DOC)
Type	Government Agency
Summary	The U.S. DOC promotes job creation, economic growth, sustainable development and improved standards of living for all Americans by working in partnership with businesses, universities, communities, and our nation's workers. The department touches the daily lives of the American people in many ways, with a wide range of responsibilities in five strategic areas, which include: trade and investment, innovation, environment, data, and operational excellence.[56]
Relevant Agencies	<ul><li>NIST: NIST is a non-regulatory federal agency within the DOC. NIST's mission is to promote U.S. innovation and industrial competitiveness by advancing measurement science, standards, and technology in ways that enhance economic security and improve our quality of life.[57]<ul><li>Information Technology Laboratory (ITL): ITL is one of the major research components of NIST. ITL accelerates the development and deployment of information and communications systems that are reliable, usable, interoperable, and secure; advances measurement science through innovations in mathematics, statistics, and computer science; and conducts research to develop the measurements and standards infrastructure for emerging information technologies and applications.[58]</li><li>Physical Measurement Laboratory (PML): The PML develops and disseminates the national standards of length, mass, force and shock, acceleration, time and frequency, electricity, temperature, humidity, pressure and vacuum, liquid and gas flow, and electromagnetic, optical, microwave, acoustic, ultrasonic, and ionizing radiation. Its activities range from fundamental measurement research through the provisioning of measurement services, standards, and data. It supports the research community in such areas as communication, defense, electronics, energy, environment, health, lighting, manufacturing, microelectronics, radiation, remote sensing, space, and transportation.[59]</li></ul></li></ul>

---

[56] Department of Commerce, *About the Department of Commerce.* Available at: https://www.commerce.gov/page/about-commerce (last accessed February 25, 2016).
[57] National Institutes of Standards and Technology, *NIST General Information.* Available at: http://www.nist.gov/public_affairs/general_information.cfm (last accessed February 25, 2016).
[58] National Institutes of Standards and Technology, *Information Technology Laboratory.* Available at: http://www.nist.gov/itl/index.cfm (last accessed February 25, 2016).
[59] National Institutes of Standards and Technology, *Physical Measurement Laboratory.* Available at: http://www.nist.gov/pml/what-we-do.cfm (last accessed February 25, 2016).

Relevant Agencies (continued)

- o Special Programs Office (SPO): The SPO fosters communication and collaboration between NIST and external communities focused on critical national needs. The focus areas include: environment, energy, forensic science, healthcare, homeland security, information technology (IT) and cybersecurity, manufacturing, and physical infrastructure. In order to do this, the SPO works closely and forges partnerships among government, military, academia, professional organizations, and the private industry to provide world-class leadership in standards and technology innovation to respond to these critical national needs.[60]

- o National Strategy for Trusted Identities in Cyberspace (NSTIC): NSTIC envisions an online environment—the "Identity Ecosystem"—that improves on the use of passwords and usernames and allows individuals and organizations to better trust one another, with minimized disclosure of personal information. The Identity Ecosystem is a user-centric online environment, a set of technologies, policies, and agreed-upon standards, that securely supports transactions ranging from anonymous to fully authenticated and from low to high value. It would include a vibrant marketplace that allows people to choose among multiple identity providers—both private and public—that would issue trusted credentials that prove identity. Key attributes of the Identity Ecosystem include privacy, convenience, efficiency, ease-of-use, security, confidence, innovation, and choice.[61]

- • National Telecommunications and Information Administration (NTIA): NTIA is an agency in the DOC that serves as the Executive Branch agency principally responsible for advising the President on telecommunications and information policies. In addition to representing the Executive Branch in both domestic and international telecommunications and information policy activities, NTIA also manages the federal use of spectrum; performs cutting-edge telecommunications research and engineering, including resolving technical telecommunications issues for the federal government and private sector; and administers infrastructure and public telecommunications facilities grants.[62]

---

[60] National Institutes of Standards and Technology, *About the Special Programs Office.* Available at: http://www.nist.gov/director/spo (last accessed February 25, 2106).

[61] National Institutes of Standards and Technology, *National Strategy for Trusted Identities in Cyberspace.* Available at: http://www.nist.gov/nstic/about-nstic.html (last accessed February 25, 2016).

[62] National Telecommunications and Information Administration, *About NTIA Standards.* Available at: http://www.ntia.doc.gov/ (last accessed February 25, 2016).

Relevant Agencies (continued)	o Institute for Telecommunication Sciences (ITS): ITS is the research and engineering laboratory of the NTIA. ITS supports such NTIA telecommunications objectives as promotion of advanced telecommunications and information infrastructure development in the U.S., enhancement of domestic competitiveness, improvement of foreign trade opportunities for U.S. telecommunications firms, and facilitation of more efficient and effective use of the radio spectrum.[63]

Standards	• Federal Information Processing Standards (FIPS) Publications (PUB)
	    o FIPS-PUB-140-2: *Security Requirements for Cryptographic Modules*
	    o FIPS-PUB-180-4: *Secure Hash Standard (SHS)*
	    o FIPS-PUB-197: *Advanced Encryption Standard (AES)*

Frameworks	• Framework for Improving Critical Infrastructure Cybersecurity[64]
Coordinated Activities	• ANSI: An MOU exists between NIST and ANSI that agrees on the need for a unified national approach to develop the best possible national and international standards. • Department of Homeland Security (DHS) Office of Interoperability and Compatibility (OIC): NIST SPO provides technical expertise to the DHS OIC. • NIEM: National Information Exchange Model
Effects on NG911	• Manages grant programs that may be used for NG911 purposes. • May affect IP networking and ESInet aspects. • Develops standards related to handling emergency data sets.
Website	http://www.commerce.gov/

---

[63] National Telecommunications and Information Administration, *Institute for Telecommunications Science Standards.* Available at: http://www.its.bldrdoc.gov/ (last accessed February 25, 2016).

[64] National Telecommunications and Information Administration, *Cybersecurity Framework.* Available at: http://www.nist.gov/cyberframework/ (last accessed February 25, 2016).

Department of Homeland Security (DHS)

Name	Department of Homeland Security (DHS)
Type	Government Agency
Summary	DHS' vision is to ensure a homeland that is safe, secure, and resilient against terrorism and other hazards. There are five core missions for DHS: • Prevent terrorism and enhance security • Secure and manage our borders • Enforce and administer our immigration laws • Safeguard and secure cyberspace • Ensure resilience to disasters[65]

Relevant Directorates	• National Protection and Programs Directorate (NPPD): The goal of NPPD is to advance the Department's risk-reduction mission. Reducing risk requires an integrated approach that encompasses both physical and virtual threats and their associated human elements.[66]
	  ○ Office of Cybersecurity and Communications (CS&C): CS&C is responsible for enhancing the security, resilience, and reliability of the nation's cyber and communications infrastructure.[67]
	    – Office of Emergency Communications (OEC): OEC supports emergency communications interoperability by offering training, tools, and workshops; by providing regional support; and by providing guidance documents and templates. These services assist OEC's stakeholders in ensuring that they have communications during steady state and emergency operations. OEC plays a key role in ensuring federal, state, local, tribal, and territorial agencies have the necessary plans, resources, and training needed to support operable and advanced interoperable emergency communications.[68]

---

[65] U.S. Department of Homeland Security, Our Mission. Available at: http://www.dhs.gov/our-mission (last accessed February 25, 2016).

[66] U.S. Department of Homeland Security, *National Protection and Programs Directorate*. Available at: http://www.dhs.gov/about-national-protection-and-programs-directorate (last accessed February 25, 2016).

[67] U.S. Department of Homeland Security, *Office of Cybersecurity and Communications*. Available at: http://www.dhs.gov/office-cybersecurity-and-communications (last accessed February 25, 2016).

[68] U.S. Department of Homeland Security, *Office of Emergency Communications*. Available at: http://www.dhs.gov/about-office-emergency-communications (last accessed February 25, 2016).

Relevant Directorates (continued)	• <u>Science & Technology Directorate (S&T)</u>: S&T is the primary research and development arm of DHS. S&T's mission is to deliver effective and innovative insight, methods and solutions for the critical needs of the Homeland Security Enterprise.[69] o <u>Office for Interoperability and Compatibility (OIC)</u>: OIC provides local, tribal, territorial, state, and federal entities with tools, technology, methodology, and guidance to improve interoperability. OIC manages a comprehensive research, development, testing, evaluation, and standards program to further enhance interoperability. [70]
Relevant Programs and Projects	• <u>Wireless Public **SAFE**ty Interoperable **COM**munications Program (SAFECOM)</u>: SAFECOM is a federal program that assists federal, state, local, tribal, and territorial public safety agencies in identifying wireless interoperable communications requirements and ensures those entities can communicate and share information to effectively respond to emergency incidents.[71] • <u>Integrated Public Alert & Warning System (IPAWS) Project</u>: The IPAWS project supports the advancement of interoperability and state-of-the-art technologies for alerts and warnings through standards development and adoption, conformity assessment, industry capability analysis, and technology evaluation. The result of these efforts will enable local, tribal, and state practitioners to provide reliable and accurate alerts and warnings to a wider public. As a result, there will be a significant reduction in the loss of life and property from all hazards.[72] • <u>Interoperability Continuum</u>: The Interoperability Continuum is designed to help the emergency response community and local, tribal, state, and federal policymakers address critical elements for success as they plan and implement interoperability solutions. These elements include governance, standard operating procedures, technology, training and exercises, and use of interoperable communications. Updated in 2008, the Continuum's technology element was divided into data and voice elements to reflect the modern path to improving interoperability via information sharing and voice communications.[73]

---

[69] U.S. Department of Homeland Security, *Science and Technology Directorate.* Available at: http://www.dhs.gov/science-and-technology/about-st (last accessed February 25, 2016).

[70] U.S. Department of Homeland Security, *Office for Interoperability and Compatibility.* Available at: http://www.dhs.gov/science-and-technology/office-interoperablity-and-compatiblity (last accessed February 25, 2016).

[71] U.S. Department of Homeland Security, SAFECOM. Available at: http://www.dhs.gov/safecom/about-safecom (last accessed February 25, 2016).

[72] Federal Emergency Management Agency, *Integrated Public Alert & Warning System (IPAWS.)* Available at: http://www.fema.gov/integrated-public-alert-warning-system (last accessed February 25, 2016).

[73] U.S. Department of Homeland Security, *SAFECOM Interoperability Continuum.* Available at: http://www.dhs.gov/sites/default/files/publications/interoperability_continuum_brochure_2.pdf (last accessed February 25, 2016).

Relevant Programs and Projects (continued)

- Voice over Internet Protocol (VoIP): The document outlines requirements, provides a detailed explanation of security threats and countermeasures that can be applied to VoIP systems to safeguard them against hazards. The Security Requirements Checklist for VoIP Systems is a part of the Protocol to ensure Component compliance. [74]

Standards

- *National Emergency Communications Plan*
- SAFECOM: *Emergency Communications Governance Guide for State, Local, Tribal, and Territorial Officials*[75]

Coordinated Activities

- OEC: The OIC, in coordination with OEC, is developing a Standard Operating Procedures (SOP) Development Guide, a Shared Channel Guide v2.0, and a brochure on plain language.
- NIEM: National Information Exchange Model

Effects on NG911

- Develops standards related to handling emergency data sets.

Website http://www.dhs.gov/

---

[74] U.S. Department of Homeland Security, *Voice over Internet Protocol.* Available at: https://www.dhs.gov/publication/voice-over-internet-protocol-voip (last accessed February 25, 2016).
[75] *Emergency Communications Governance Guide for State, Local, Tribal, and Territorial Officials.* Available at: https://www.dhs.gov/publication/governance-documents (last accessed February 25, 2016).

Department of Justice (DOJ)

Name	Department of Justice (DOJ)
Type	Government Agency
Summary	DOJ's mission is to enforce the law and defend the interests of the U.S. according to the law; to ensure public safety against threats foreign and domestic; to provide federal leadership in preventing and controlling crime; to seek just punishment for those guilty of unlawful behavior; and to ensure fair and impartial administration of justice for all Americans. [76]
Relevant Directorates	• Office of Justice Programs (OJP): OJP's mission is to increase public safety and improve the fair administration of justice across America through innovative leadership and programs. [77]
Relevant Bureaus and Offices	• Bureau of Justice Assistance (BJA): BJA's mission is to provide leadership and services in grant administration and criminal justice policy development to support local, state, and tribal justice strategies to achieve safer communities. BJA supports programs and initiatives in the areas of law enforcement, justice information sharing, countering terrorism, managing offenders, combating drug crime and abuse, adjudication, advancing tribal justice, crime prevention, protecting vulnerable populations, and capacity building. [78]
Standards	• *Criminal Justice Information Services (CJIS) Security Policy*
Coordinated Activities	• NIEM: National Information Exchange Model
Effects on NG911	• Develops standards related to handling emergency data sets, specifically pertaining to interoperability for data sharing.
Website	http://www.justice.gov/

---

[76] United States Department of Justice, *About DOJ*. Available at: http://www.justice.gov/about/about.html (last accessed February 25, 2016)

[77] Office of Justice Programs, *Mission and Vision*. Available at: http://www.ojp.usdoj.gov/about/mission.htm (last accessed February 25, 2016).

[78] Office of Justice Programs, *About the Bureau of Justice Assistance*. Available at: https://www.bja.gov/About/index.html (last accessed February 25, 2016).

Department of Transportation (USDOT)

Name	Department of Transportation (USDOT)
Type	Government Agency
Summary	USDOT serves the U.S. by ensuring a fast, safe, efficient, accessible, and convenient transportation system that meets our vital national interests and enhances the quality of life of the American people, today and into the future. [79]
Relevant Administrations	• NHTSA: NHTSA was established by the Highway Safety Act of 1970 and is dedicated to achieving the highest standards of excellence in motor vehicle and highway safety. It works daily to help prevent crashes and their attendant costs, both human and financial.[80] • Office of the Assistant Secretary for Research and Technology (OST-R): Comprises all of the program offices, statistics, and research activities previously administered by RITA. The mission of OST-R is to transform transportation by expanding the base knowledge to make America's transportation system safer, more competitive and sustainable. In an effort to accomplish this goal, OST-R is responsible to: advance innovation, technology development, and breakthrough knowledge; conduct research and facilitate multimodal research collaboration; foster technology transfer through partnerships within the Department and with other partners; provide useful information and statistics to decision-makers as they debate policies; and develop a highly skilled interdisciplinary transportation workforce for the nation.[81]

---

[79] United States Department of Transportation, *About DOT*. Available at: http://www.dot.gov/about.html (last accessed February 25, 2016).

[80] National Highway Traffic Safety Administration, *About NHTSA*. Available at: http://www.nhtsa.gov/About (last accessed February 25, 2016).

[81] Office of the Assistant Secretary for Research and Technology, About OST-R. Available at: http://www.rita.dot.gov/about_rita (last accessed February 25, 2016).

Relevant Administrations (continued)	o Intelligent Transportation Systems Joint Program Office (ITS JPO): The ITS JPO program focuses on intelligent vehicles, intelligent infrastructure, and the creation of an intelligent transportation system through integration with and between these two components. The federal ITS program supports the overall advancement of ITS through investments in major initiatives, exploratory studies, and a deployment support program. Increasingly, the federal investments are directed at targets of opportunity—major initiatives—that have the potential for significant payoff in improving safety, mobility, and productivity.[82]
	o Transportation Safety Advancement Group (TSAG): The TSAG serves an important function on behalf of the USDOT, OST-R, and the ITS JPO. Through its members and allied stakeholder groups, TSAG identifies surface transportation-based technologies and applications, and promotes a national dialogue regarding public safety practitioners' first-hand experiences, corresponding best practices, and lessons learned. [83]
Relevant Programs and Projects	• Next Generation 911 (NG911) Initiative: The nation's current 911 system is designed around telephone technology and cannot handle the text, data, images, and video that are increasingly common in personal communications and critical to future transportation safety and mobility advances. The NG911 Initiative has established the foundation for public emergency communications services in a wireless mobile society.[84]
	• National 911 Program: The National 911 Program, in coordinating the efforts of states, technology providers, public safety officials, 911 professionals and other groups, seeks to ensure a smooth, reliable, and cost-effective transition to a 911 system that takes advantage of new communications technologies to enhance public safety nationwide.[85]
Coordinated Activities	• ETSI: A memorandum of cooperation exists between USDOT/OST-R/ITS and ETSI • FCC CSRIC: Seminars and coordination
Websites	http://www.dot.gov/ http://911.gov/

---

[82] Intelligent Transportation Systems, *ITS Overview.* Available at: http://www.its.dot.gov/factsheets/overview_factsheet.htm (last accessed February 25, 2016).

[83] Transportation Safety Advancement Group, *About TSAG.* Available at: http://www.tsag-its.org/whatistsag.php (last accessed February 25, 2016).

[84] Research and Innovation Technology Administration, *Next Generation 911.* Available at: http://www.its.dot.gov/ng911/index.htm (last accessed February 25, 2016).

[85] 911.gov, *About The National 911 Program.* Available at: http://www.911.gov/about_national_911program.html (last accessed February 25, 2016).

European Telecommunications Standards Institute (ETSI)

Name　　　　　European Telecommunications Standards Institute (ETSI)

Type　　　　　Regional Standards Organization

Summary　　　ETSI is an independent, not-for-profit organization that produces globally applicable standards for ICT, including fixed, mobile, radio, converged, broadcast, and Internet technologies.[86]

Relevant Committees and Other Bodies

- EMTEL—Emergency Communications: EMTEL addresses a broad spectrum of issues related to the use of telecommunications services in emergency situations.[87]
- TISPAN - Telecommunications & Internet Converged Services & Protocols for Advanced Networks: ETSI TISPAN has been the key standardization body in creating the Next Generation Networking (NGN) specifications.[88]
- Next Generation Protocols (NGP) Industry Specification Group: NGP is looking at evolving communications and networking protocols to provide the scale, security, mobility and ease of deployment required for the connected society of the 21st century.[89]

Standards

- ETSI SR 002 777: *Emergency Communications (EMTEL); Test/verification procedure for emergency calls*
- ETSI TS 101 470: *Emergency Communications (EMTEL); Total Conversion Access to Emergency Services*
- ETSI TS 102 164: *Telecommunications and converged Services and Protocols for Advanced Networking (TISPAN); Emergency Location Protocols*
- ETSI TR 102 180: *Emergency Communications (EMTEL); Basis of requirements for communication of individuals with authorities/organizations in case of distress (Emergency call handling)*
- ETSI TS 102 424: *Telecommunications and Internet converged Services and Protocols for Advanced Networking (TISPAN); Requirements of the NGN network to support Emergency Communication from Citizen to Authority*

---

[86] European Telecommunications Standards Institute, *Introduction*. Available at: http://www.etsi.org/about (last accessed February 25, 2016).

[87] European Telecommunications Standards Institute, *EMTEL Overview*. Available at: http://www.emtel.etsi.org/overview.htm (last accessed February 25, 2016).

[88] European Telecommunications Standards Institute, *Telecoms & Internet Services & Protocols for Advanced Network. Overview*. Available at: http://www.etsi.org/tispan/ (last accessed February 25, 2016).

[89] European Telecommunications Standards Institute, *Next Generation Protocols (NGP)*. Available at: http://www.etsi.org/technologies-clusters/technologies/next-generation-protocols (last accessed February 25, 2016).

Standards (continued)	• ETSI TR 103 170: *Emergency Communications (EMTEL); Total Conversation Access to Emergency Services*
	• ETSI TS 123 167: *Universal Mobile Telecommunications System (UMTS); LTE; IP Multimedia Subsystem (IMS) emergency sessions*
	• ETSI TS 182 009: *Telecommunications and Internet converged Services and Protocols for Advanced Networking (TISPAN); NGN Architecture to support emergency communication from citizen to authority*
	• ETSI TS 183 036: *Telecommunications and Internet converged Services and Protocols for Advanced Networking (TISPAN); ISDN/SIP interworking; Protocol specification*
	• ETSI TS 187 001: *Telecommunications and Internet converged Services and Protocols for Advanced Networking (TISPAN); NGN SECurity (SEC); Requirements*
	• ETSI TR 187 002: *Telecommunications and Internet converged Services and Protocols for Advanced Networking (TISPAN); TISPAN NGN Security (NGN_SEC); Threat, Vulnerability and Risk Analysis*
	• ETSI TS 187 003: *Telecommunications and Internet converged Services and Protocols for Advanced Networking (TISPAN); NGN Security; Security Architecture*
	• ETSI TS 187 005: *Telecommunications and Internet converged Services and Protocols for Advanced Networking (TISPAN); NGN Lawful Interception; Stage 1 and Stage 2 definition*
	• ETSI ES 203 178: *Functional architecture to support European requirements on emergency caller location determination and transport*
	• ETSI ES 282 007: *Telecommunications and Protocols for Advanced Networking (TISPAN); IP Multimedia Subsystem (IMS); Functional architecture*
Coordinated Activities	• 3GPP
	• USDOT: A memorandum of cooperation exists between USDOT/OST-R/ITS and ETSI
Effects on NG911	• Develops standards that enable text and multimedia transmission from the caller to the NG911 system (transport of data).
	• Develops standards adhered to by OSP's network and applications services for emergency calling.
	• Supports location requirements and standards.
Website	http://www.etsi.org/

Federal Communications Commission (FCC)

Name	Federal Communications Commission (FCC)
Type	Government Agency
Summary	The FCC is an independent United States government agency charged with regulating interstate and international communications by radio, television, wire, satellite, and cable. [90]
Relevant Bureaus	• Public Safety and Homeland Security Bureau (PSHSB): The FCC's PSHSB is responsible for developing, recommending, and administering the agency's policies pertaining to public safety communications issues. These policies include 911 and E911; operability and interoperability of public safety communications; communications infrastructure protection and disaster response; and network security and reliability. PSHSB also serves as a clearinghouse for public safety communications information and emergency response issues.[91]
Relevant Advisory Committees	• CSRIC: CSRIC's mission is to provide recommendations to the FCC to ensure, among other things, optimal security and reliability of communications systems, including telecommunications, media, and public safety.[92] CSRIC councils are appointed by the Chairman of the FCC and are typically chartered for two years. The following are CSRIC councils and associated working groups relevant to NG911: ○ CSRIC II Charter Term: March 19, 2009–March 18, 2011 − Working Group 2A–Cyber Security Best Practices: While a large body of cyber security best practices were previously created by the Network Reliability and Interoperability Council (NRIC), many years have passed and the state of the art in cyber security has advanced rapidly. This Working Group took a fresh look at cyber security best practices, including all segments of the communications industry and public safety communities.

[90]FCC, *About the FCC.* Available at: http://www.fcc.gov/aboutus.html (last accessed February 25, 2016).

[91] FCC, *Public Safety and Homeland Security Bureau, About Us.* Available at: https://www.fcc.gov/public-safety-and-homeland-security (last accessed February 25, 2016).

[92]FCC, *Communications Security, Reliability and Interoperability Council.* Available at: https://www.fcc.gov/about-fcc/advisory-committees/general/communications-security-reliability-and (last accessed February 25, 2016).

**Relevant
Advisory
Committees
(continued)**

- Working Group 4A–Best Practices for Reliable 911 and E911: Investigated and evaluated currently available 911-related VoIP standards and best practices related to E911 for completeness and to identify any gaps, including challenges related to implementation of such standards by VoIP providers within the E911 system. The Working Group evaluated and recommended to CSRIC how to resolve incomplete work and gaps, identified and recommended what groups should perform that work (including the option of the CSRIC Working Group doing so), and recommended to CSRIC an appropriate work schedule.[93]
- Working Group 4B–Transition to NG911: Building on the work of Working Group 4A, the group determined what changes or additions in 911-related VoIP standards and best practices are required for the evolution of IP-based originating service providers to the IP-based NG911 system environment, both during the transition from E911 to NG911 and as identifiable for the longer term all-IP NG911 environment. The Working Group considered technical issues as well as operational and funding challenges for PSAPs in an NG911 environment. The Working Group also considered ways that NG911 architectures and technologies can improve 911 access for people with disabilities and non-English speaking communities.
- Working Group 4C–Technical Options for E911 Location Accuracy: The group examined E911/Public Safety location technologies in use today, identifying current performance and limitations for use in NG Public Safety Applications. They also examined emerging E911/Public Safety location technologies and recommended options to CSRIC for improvement of E911 location accuracy including implementation timelines.

---

[93] https://transition.fcc.gov/pshs/advisory/csric/wg-descriptions.pdf

Relevant Advisory Committees (continued)

- o <u>CSRIC III</u> Charter Term: March 19, 2011–March 18, 2013
  - – Working Group 1–NG 9-1-1: Recommended additional standards work needed to enable NG911 network architecture, particularly those related to NENA's i3 standard, and related standards needed from other organizations such as the Internet Engineering Task Force (IETF), 3GPP, and ATIS. The Working Group identified gaps in NG911 network architecture standards and labeled them.[94]
  - – Working Group 3–E911 Location Accuracy: Examined E911/public safety indoor and outdoor location technologies in use today, identifying current performance and limitations for use in next generation public safety applications. More specifically, the Working Group examined emerging E911/public safety location technologies and recommended options to CSRIC for improvement of E911 location accuracy, including implementation timelines.[95]
  - – Working Group 8–E911 Best Practices: Reviewed the existing CSRIC/NRIC 911 best practices and recommended ways to improve them, accounting for the passage of time, technology changes, operational factors, and any identified gaps. As part of this effort, the Working Group provided recommendations regarding the creation of two new, non-industry best practice categories: (i) PSAP and (ii) 911 Consumer. The Working Group also provided recommendations regarding how to better engage PSAPs in the best practice process.[96]

---

[94] Communications Security, Reliability, and Interoperability Council, *CSRIC III Working Group Descriptions*. Available at: http://transition.fcc.gov/pshs/advisory/csric3/wg-descriptions_2-28-12.pdf (last accessed February 26, 2016).

[95] Communications Security, Reliability, and Interoperability Council, *CSRIC III Working Group Descriptions*. Available at: http://transition.fcc.gov/pshs/advisory/csric3/wg-descriptions_2-28-12.pdf (last accessed February 26, 2016).

[96] Communications Security, Reliability, and Interoperability Council, *CSRIC III Working Group Descriptions*. Available at: http://transition.fcc.gov/pshs/advisory/csric3/wg-descriptions_2-28-12.pdf (last accessed February 26, 2016).

Relevant Advisory Committees (continued)

- o CSRIC IV Charter Term: March 19, 2013–March 18, 2015
  - – Working Group 1–NG911: The Working Group reported on the technical feasibility for wireless carriers to include E911 Phase 2 location accuracy and information in texts sent to 911, and made recommendations for including enhanced location information in texts to 911. In addition, the Working Group recommended best practices— including testing and trialing—operational procedures, and security requirements that wireless carriers, PSAPs, and third-party service providers should follow in provisioning PSAP requests for text-to-911 service.[97]
  - – Working Group 4–Cyber Security Best Practices: The Working Group developed voluntary mechanisms to provide macro-level assurance to the FCC and the public that communications providers are taking the necessary corporate and operational measures to manage cyber security risks across the enterprise. The mechanisms demonstrate how communications providers are reducing cyber security risks through the application of the NIST Cybersecurity Framework, or an equivalent construct.[98]
- o CSRIC V Charter Term: March 19, 2015–March 18, 2017
  - – Working Group 1–Evolving 911 Services: The Working Group is reviewing public safety and industry best practices and SOPs for rerouting 911 calls between PSAPs resulting from the use of cell sectors for routing purposes, and where necessary identify gaps and make recommendations towards mitigating PSAP call transfers and optimizing rerouting best practices (Task1). The group is also studying and making recommendations on the architectural, technical, operational standards, and cyber security requirements of location-based routing that uses longitude and latitude information or other location identification methods (when available) to determine and route a 911 call to the nearest appropriate PSAP (Task 2).[99]

---

[97] Communications Security, Reliability, and Interoperability Council, *CSRIC IV Working Group Descriptions*. Available at: http://transition.fcc.gov/bureaus/pshs/advisory/csric4/CSRIC_IV_Working_Group_Descriptions_12_31_13.pdf (last accessed February 26, 2016).

[98] Communications Security, Reliability, and Interoperability Council, CSRIC IV Working Group Descriptions. Available at: https://transition.fcc.gov/bureaus/pshs/advisory/csric4/CSRIC%20IV%20Working%20Group%20Descriptions%201 0%2023%2014.pdf (last accessed February 26, 2016).

[99] Communications Security, Reliability, and Interoperability Council, CSRIC V Working Group Descriptions. Available at: https://transition.fcc.gov/bureaus/pshs/advisory/csric5/Working_GroupCSRICV_110515.pdf (last accessed March 1, 2016).

Relevant Advisory Committees (continued)

- Working Group 8–Priority Services: Priority communications over commercial networks during a national emergency remains as essential today to responders and national security personnel as in decades past. However, commercial communications networks are increasingly relying on packet-based technology and retiring Time Division Multiplexing (TDM) technology. The Federal government is losing priority capabilities throughout this transition, as voice priority services rely on wireline TDM, which will eventually be replaced by IP-based infrastructure. Lack of priority communications services on packet-based systems could jeopardize national security or domestic incident response. This Working Group is assessing how priority services programs can take advantage of packet-based technologies and will recommend protocols that can be used to ensure priority communications upon retirement of TDM.

- Emergency Response Interoperability Center (ERIC): The mission of ERIC is to establish a technical and operational framework that will ensure nationwide operability and interoperability in deployment and operation of the 700 megahertz (MHz) public safety broadband wireless network. ERIC will adopt, implement, and coordinate interoperability regulations, license requirements, grant conditions, and technical standards. DHS, NIST, DOJ, and DOC contribute to ERIC's functions.[100]

- Emergency Access Advisory Committee (EAAC): The EAAC Charter expired in July 2013. EAAC was chartered to determine the most effective and efficient technologies and methods by which to enable equal access to emergency services by individuals with disabilities as part of the nation's migration to NG911, and to make recommendations to the Commission on how to achieve those effective and efficient technologies and methods.[101]

- NRIC: NRIC was an advisory council, chartered by the FCC to partner with the FCC, the communications industry, and public safety to facilitate enhancement of emergency communications networks, homeland security, and best practices across the burgeoning telecommunications industry. There were seven assemblies of NRIC since 1992. NRIC is no longer active and has been superseded by CSRIC within the FCC. The documents from NRIC can be accessed from the CSRIC website.

---

[100] Federal Communication Commission, Public Safety and Homeland Security Bureau, *Emergency Response Interoperability Center (ERIC)*. Available at: http://transition.fcc.gov/pshs/eric.html (last accessed February 26, 2016).

[101] Federal Communication Commission, Emergency Access Advisory Committee (EAAC). Available at: http://www.fcc.gov/encyclopedia/emergency-access-advisory-committee-eaac (last accessed February 26, 2016).

Relevant Advisory Committees (continued)	• <u>Task Force on Optimal PSAP Architecture (TFOPA)</u>: TFOPA is a federal advisory committee that will provide recommendations to the Commission regarding actions that PSAPs can take to optimize their security, operations, and funding as they migrate to NG911.[102]
Standards	• The <u>CSRIC Best Practices Search Tool</u> allows you to search CSRIC's collection of Best Practices using a variety of criteria including Network Type, Industry Role, Keywords, Priority Levels, and BP Number.[103] • CSRIC IV Working Group 1 Next Generation 9-1-1 Task 1 Subtask 1: *Investigation into Location Improvements for Interim SMS (Text) to 9-1-1* • CSRIC IV Working Group 1 Next Generation 9-1-1 Task 1 Subtask 2: *PSAP Requests for Service for Interim SMS Text-to-9-1-1* • CSRIC IV Working Group 1 Next Generation 9-1-1 Task 2: *Location Accuracy and Testing for Voice-over-LTE Networks* • CSRIC IV Working Group 1 Next Generation 911 Task 3: *Specification for Indoor Location Accuracy Test Bed* • CSRIC IV Working Group 4: *Cyber Security Best Practices*
Website	http://www.fcc.gov/

---

[102] Federal Communication Commission, Task Force on Optimal Public Safety Answering Point Architecture (TFOPA). Available at: https://www.fcc.gov/about-fcc/advisory-committees/general/task-force-optimal-public-safety-answering-point (last accessed February 25, 2016).

[103] Federal Communication Commission, Public Safety and Homeland Security Bureau, CSRIC Best Practices. Available at: https://www.fcc.gov/nors/outage/bestpractice/BestPractice.cfm (last accessed February 25, 2016).

Federal Geographic Data Committee (FGDC)

Name   Federal Geographic Data Committee (FGDC)

Type   Interagency Committee

Summary   FGDC is an interagency committee that promotes the coordinated development, use, sharing, and dissemination of geospatial data on a national basis. This nationwide data publishing effort is known as the National Spatial Data Infrastructure (NSDI). The NSDI is a physical, organizational, and virtual network designed to enable the development and sharing of this nation's digital geographic information resources. FGDC activities are administered through the FGDC Secretariat, hosted by the U.S. Geological Survey. [104]

Relevant Agencies
- FGDC Structure and Federal Agency and Bureau Representation: In accordance with Office of Management and Budget (OMB) Circular A-16, the FGDC is chaired by the Secretary of the Interior with the Deputy Director for Management, OMB as Vice-Chair.[105]

Standards
- FGDC-STD-016-2011: *United States Thoroughfare, Landmark, and Postal Address Data Standard*

Coordinated Activities
- OMB and the U.S. Congress set policy for federal agencies. The FGDC, a federal interagency coordinating committee, is guided by those policies in the design of programs, activities, and technologies. The FGDC sets geospatial information policy in harmony with overall information policy. The FGDC Secretariat engages in on-going strategic planning to ensure continued investment of resources in high-value programs, activities and technologies. [106]

Effects on NG911
- Develops standards pertaining to interoperability for data sharing.

Website   http://www.fgdc.gov/

---

[104] Federal Geographic Data Committee. Available at: http://www.fgdc.gov/ (last accessed February 26, 2016).
[105] Federal Geographic Data Committee Structure. Available at: http://www.fgdc.gov/participation (last accessed February 26, 2016).
[106]Federal Geographic Data Committee Policy and Planning. Available at: http://www.fgdc.gov/policyandplanning (last accessed February 26, 2016).

Information Security Forum (ISF)

Name   Information Security Forum (ISF)

Type   Global Information Systems Security and Risk Management Organization

Summary   ISF, an independent, not-for-profit association of leading organizations from around the world, investigates, clarifies and resolves key issues in cyber, information security, and risk management. The ISF develops best practice methodologies, processes, and solutions. The ISF's membership includes some of the world's major corporations, public sector bodies, and government departments. The ISF has a range of products and services available to members and non-members including research reports, tools, methodologies, and free webinars.[107]

Relevant Projects
- Risk Management Tool: The ISF's Risk Manager is designed to help organizations develop impact assessments, threat and vulnerability assessments, and evaluate and select controls to help mitigate threats to the organization.
- Benchmark: This tool help organizations manage and control information risk throughout their enterprise. The Benchmark allows comparison of organizational security arrangements against seven internationally recognized standards:
  - ISF *Standard of Good Practice for Information Security*
  - NIST *Cyber Security Framework*
  - The SANS Top 20 Critical Security Controls for Effective Cyber Defense
  - Payment Card Industry Data Security Standard (PCI DSS) version 3.1
  - ISO/IEC 27002: 2013
  - COBIT 5 for Information Security
  - ISO/IEC 27002: 2005
- ISF Webinar Programme: The webinars provide organizations with an opportunity to find out more about research into key topics such as identifying and managing current and emerging threats. This includes content such as the unintended consequences of state intervention, big data, mobile applications and the Internet of Things (IoT), cybercrime and the growing skills gap in the information security industry.

---

[107] ISF, *About ISF.* Available at: https://www.securityforum.org/about (last accessed February 26, 2016).

Standards   Standard of Good Practice for Information Security: Standard for IT
                     personnel to manage information risk, enable compliance with ISO/IEC
                     27002:2013, COBIT 5 for Information Security and the SANS Top 20 Critical
                     Security Controls. It also provides organizations with detailed controls
                     which can help comply with the NIST *Cyber Security Framework*.

Websites   https://www.securityforum.org

Institute of Electrical and Electronics Engineers (IEEE)

Name	Institute of Electrical and Electronics Engineers (IEEE)
Type	Professional Association
Summary	IEEE is the world's largest professional association with the core purpose to advance technological innovation and excellence for the benefit of humanity. IEEE and its members support a global community through a variety of activities including the development of technology standards. [108]
Relevant Committees	• IEEE 802 LAN/MAN Standards Committee: The IEEE 802 Working Group and Study Groups address Local Area Network (LAN)/Metropolitan Area Network (MAN) Standards Committee develops LAN standards and MAN standards.[109] ○ IEEE 802.1 Working Group: The IEEE 802.1 Working Group is chartered to concern itself with and develop standards and recommend practices in the following area: 802 LANs, MANs, and other wide area networks (WANs), 802 Security, 802 overall network management, and protocol layers above the Media Access Control (MAC) and Logical Link Control (LLC) layers. The 802.1 Working Group has four active task groups: Interworking, Security, Audio/Video Bridging and Data Center Bridging.[110] ○ IEEE 802.11 Wireless Local Area Networks Working Group: The IEEE 802.11 Working Group develops standards and recommended practices to support development and deployment of wireless local area networks (WLANs).[111] ○ IEEE 802.16 Broadband Wireless Access Standards Working Group: The IEEE 802.16 Working Group develops standards and recommended practices to support development and deployment of broadband wireless MANs.[112]

---

[108]IEEE, *About IEEE.* Available at: http://www.ieee.org/about/index.html (last accessed February 26, 2016).

[109] IEEE, *IEEE 802 LAN / MAN Standards Committee.* Available at: http://grouper.ieee.org/groups/802/index.shtml (last accessed February 26, 2016).

[110] IEEE, *IEEE 802.1 Working Group.* Available at: http://www.ieee802.org/1/ (last accessed February 26, 2016).

[111] IEEE, *IEEE 802.11 Wireless Local Area Networks.* Available at: http://www.ieee802.org/11/ (last accessed February 26, 2016).

[112] IEEE, *IEEE 802.16 Working Group on Broadband Wireless Access Standards.* Available at: http://www.ieee802.org/16/ (last accessed February 26, 2016).

Relevant Committees (continued)	○ IEEE 802.23 Emergency Services Working Group: The IEEE 802.23 Working Group developed standards and recommended practices to support a framework that provides consistent access and data facilitating compliance with applicable civil authority requirements for communications systems that include IEEE 802 networks.[113] Due to lack of participation, this Working Group is no longer active and was disbanded in June 2011.
Standards	• IEEE 802.1AB-2009: *Station and Media Access Control Connectivity Discovery* • IEEE 802.1AC: *Media Access Control (MAC) Services Definition* • IEEE 802.11-2012: *Wireless LAN Medium Access Control (MAC) and Physical Layer (PHY) Specifications* • IEEE 802.16-2012: *Air Interface for Broadband Wireless Access Systems* • IEEE 802.23: *Emergency Services for Internet Protocol (IP) Based Citizen to Authority Communications,* Draft • IEEE 1512-2006: *Common Incident Management Message Sets for Use by Emergency Management Centers* • IEEE 1903-2011: *Functional Architecture of Next Generation Service Overlay Networks*
Coordinated Activities	• WiMAX Forum • 3GPP • IETF • ANSI: IEEE is an ANSI-accredited SDO.
Website	http://www.ieee.org/

---

[113] IEEE, *IEEE 802.23 Emergency Services Working Group.* Available at: http://www.ieee802.org/23/ (last accessed February 26, 2016).

Internet Engineering Task Force (IETF)

Name Internet Engineering Task Force (IETF)

Type International Standards Organization—Industry (Networking)

Summary IETF's mission is to produce high quality, relevant technical and engineering documents that influence the way people design, use, and manage the Internet, in such a way as to make the Internet work better. These documents include protocol standards, current best practices, and informational documents of various kinds.[114]

Relevant Working Groups
- Emergency Context Resolution with Internet Technologies (ECRIT): In a number of areas, the public switched telephone network (PSTN) has been configured to recognize an explicitly specified number as a call for emergency services. These numbers (e.g., 911, 112) relate to an emergency service context and depend on a broad, regional configuration of service contact methods and a geographically constrained context of service delivery. Successful delivery of an emergency service call within those systems requires both an association of the physical location of the originator with an appropriate emergency service center and call routing to deliver the call to the center. Calls placed using Internet technologies do not use the same systems to achieve those goals, and the common use of overlay networks and tunnels (either as virtual private networks [VPNs] or for mobility) makes meeting them more challenging. There are, however, Internet technologies available to describe location and to manage call routing. This Working Group will describe when these may be appropriate and how they can be used, and is considering emergency services calls that might be made by any user of the Internet.[115]

---

[114] IETF, *Mission Statement.* Available at: http://www.ietf.org/about/mission.html (last accessed February 26, 2016).

[115]IETF, *Emergency Context Resolution with Internet Technology (ECRIT).* Available at: http://datatracker.ietf.org/wg/ecrit/charter/ (last accessed February 26, 2016).

Relevant Working Groups (continued)

- Geographic Location/Privacy (GEOPRIV): As of November 2014, this Working Group is listed as a "concluded." IETF recognized that many applications are emerging that require geographic and civic location information about resources and entities, and that the representation and transmission of that information had significant privacy and security implications. It has created a suite of protocols that allows such applications to represent and transmit such location objects and to allow users to express policies on how these representations are exposed and used. GEOPRIV was chartered to continue to develop and refine representations of location in Internet protocols and to analyze the authorization, integrity, and privacy requirements that must be met when these representations of location are created, stored, and used. The group created and refined mechanisms for the transmission of these representations to address the requirements that have been identified.[116]

Standards

- Request for Comment (RFC) 2328: *OSPF Version 2*
- RFC 2474: *Definition of the Differentiated Services Field (DS Field) in the IPv4 and IPv6 Headers*
- RFC 2475: *An Architecture for Differentiated Services*
- RFC 3261: *SIP: Session Initiation Protocol*
- RFC 3262: *Reliability of Provisional Responses in Session Initiation Protocol (SIP)*
- RFC 3263: *Session Initiation Protocol (SIP): Locating SIP Servers*
- RFC 3264: *An Offer/Answer Model with Session Description Protocol (SDP)*
- RFC 3265: *Session Initiation Protocol (SIP)-Specific Event Notification*
- RFC 3411: *An Architecture for Describing Simple Network Management Protocol (SNMP) Management Frameworks*
- RFC 3412: *Message Processing and Dispatching for the Simple Network Management Protocol (SNMP)*
- RFC 3413: *Simple Network Management Protocol (SNMP) Applications*
- RFC 3414: *User-based Security Model (USM) for version 3 of the Simple Network Management Protocol (SNMPv3)*
- RFC 3415: *View-based Access Control Model (VACM) for the Simple Network Management Protocol (SNMP)*
- RFC 3416: *Version 2 of the Protocol Operations for the Simple Network Management Protocol (SNMP)*
- RFC 3417: *Transport Mappings for the Simple Network Management Protocol (SNMP)*
- RFC 3418: *Management Information Base (MIB) for the Simple Network Management Protocol (SNMP)*
- RFC 3550: *RTP: A Transport Protocol for Real-Time Applications*
- RFC 3693: *Geopriv Requirements*

---

[116] IETF, *Geographic Location / Privacy (geopriv)*. Available at: http://datatracker.ietf.org/wg/geopriv/charter/ (last accessed February 26, 2016).

Standards (continued)

- RFC 3856: *A Presence Event Package for the Session Initiation Protocol (SIP)*
- RFC 3863: *Presence Information Data Format (PIDF)*
- RFC 3966: *The tel URI for Telephone Numbers*
- RFC 3986: *Uniform Resource Identifier (URI): Generic Syntax*
- RFC 4079: *A Presence Architecture for the Distribution of GEOPRIV Location Objects*
- RFC 4119: *A Presence-based GEOPRIV Location Object Format*
- RFC 4271: *A Border Gateway Protocol 4 (BGP-4)*
- RFC 4975: *The Message Session Relay Protocol (MSRP)*
- RFC 4976: *Relay Extensions for the Message Sessions Relay Protocol (MSRP)*
- RFC 5069: *Security Threats and Requirements for Emergency Call Marking and Mapping*
- RFC 5139: *Revised Civic Location Format for Presence Information Data Format Location Object (PIDF-LO)*
- RFC 5222: *LoST: A Location-to-Service Translation Protocol* (updated by RFC 6848)
- RFC 5223: *Discovering Location-to-Service Translation (LoST) Servers Using the Dynamic Host Configuration Protocol (DHCP)*
- RFC 5246: *The Transport Layer Security (TLS) Protocol Version 1.2* (update in progress)
- RFC 5340: *OSPF for IPv6*
- RFC 5411: *A Hitchhiker's Guide to the Session Initiation Protocol (SIP)*
- RFC 5491: *GEOPRIV Presence Information Data Format (PIDF-LO) Usage Clarification, Considerations, and Recommendations*
- RFC 5582: *Location-to-URL Mapping Architecture and Framework*
- RFC 5880: *Bidirectional Forwarding Detection (BFD)*
- RFC 5881: *Bidirectional Forwarding Detection (BFD) for IPv4 and IPv6 (Single Hop)*
- RFC 5882: *Generic Application of Bidirectional Forwarding Detection (BFD)*
- RFC 5985: *HTTP-Enabled Location Delivery (HELD)*
- RFC 6135: *An Alternative Connection Model for the Message Session Relay Protocol (MSRP)*
- RFC 6155: *Use of Device Identity in HTTP-Enabled Location Delivery (HELD)*
- RFC 6442: *Location Conveyance for the Session Initiation Protocol*
- RFC 6446: *Session Initiation Protocol (SIP) Event Notification Extension for Notification Rate Control*
- RFC 6447: *Filtering Location Notifications in the Session Initiation Protocol (SIP)*
- RFC 6714: *Connection Establishment for Media Anchoring (CEMA) for the Message Session Relay Protocol (MSRP)*
- RFC 6739: *Synchronizing Service Boundaries and <mapping> Elements Based on the Location-to-Service Translation (LoST) Protocol*
- RFC 6753: *A Location Dereference Protocol Using HTTP-Enabled Location Delivery (HELD)*
- RFC 6772: *Geolocation Policy: A Document Format for Expressing Privacy Preferences for Location Information*

Standards (continued)	• RFC 6739: *Synchronizing Service Boundaries and <mapping> Elements Based on the Location-to-Service Translation (LoST) Protocol*
	• RFC 6848: *Specifying Civic Address Extensions in the Presence Information Data Format Location Object (PIDF-LO)*
	• RFC 6881: *Best Current Practice for Communications Services in Support of Emergency Calling*
	• RFC 6915: *Flow Identity Extension for HTTP-Enabled Location Delivery (HELD)*
	• RFC 7035: *Relative Location Representation*
	• RFC 7090: *Public Safety Answering Point (PSAP) Callback*
	• RFC 7105: *Using Device-Provided Location-Related Measurements in Location Configuration Protocols*
	• RFC 7163: *URN for Country-Specific Emergency Services*
	• RFC 7199: *Location Configuration Extensions for Policy Management*
	• RFC 7216: *Location Information Server (LIS) Discovery Using IP Addresses and Reverse DNS*
	• RFC 7378: *Trustworthy Location*
	• RFC 7406: *Extensions to the Emergency Services Architecture for Dealing With Unauthenticated and Unauthorized Devices*
	• RFC 7459: *Representation of Uncertainty and Confidence in the Presence Information Data Format Location Object (PIDF-LO)*
	• RFC 7701: *Multi-party Chat Using the Message Session Relay Protocol (MSRP)*
	• Internet Draft: *Additional Data Related to an Emergency Call*
	• Internet Draft: *Data Only Emergency Calls*
	• Internet Draft: *A Routing Request Extension for the HELD Protocol*
	• Internet Draft: *A LoST extension to return complete and similar location info*
	• Internet Draft: *Next-Generation Vehicle-Initiated Emergency Calls*
	• Internet Draft: *MSRP over Data Channels*
	• Internet Draft: *The WebSocket Protocol as a Transport for the Message Session Relay Protocol (MSRP)*
	• Internet Draft - EXPIRED: *Overview for MSRP Recording based on SIPREC*
Coordinated Activities	• ETSI EMTEL
	• NENA
Effects on NG911	• Develops standards that enable text and multimedia transmission from the caller to the NG911 system (transport of data).
Website	http://www.ietf.org/

International Academies of Emergency Dispatch (IAED)

Name	International Academies of Emergency Dispatch (IAED)
Type	Professional Organization
Summary	IAED's mission is to advance and support the public safety emergency telecommunications professional and ensure that citizens in need of emergency, health, and social services are matched safely, quickly, and effectively with the most appropriate resource.[117]
Certifications	<ul><li>ETC: Emergency Telecommunicator Certification</li><li>EMD: Emergency Medical Dispatch Certification</li><li>EFD: Emergency Fire Dispatch Certification</li><li>EPD: Emergency Police Dispatch Certification</li><li>ED-Q: Quality Improvement Certification</li><li>CMM: Communication Center Manager</li><li>Executive Certification Course</li></ul>
Effects on NG911	<ul><li>May drive requirements based on call-handling protocols.</li></ul>
Website	http://www.emergencydispatch.org/

---

[117] IAED, *Organization.* Available at: http://www.emergencydispatch.org/Organization (last accessed February 26, 2016).

International Organization of Standardization (ISO)

Name	International Organization of Standardization (ISO)
Type	International Standards Organization
Summary	ISO is the world's largest developer and publisher of international standards. ISO is a network of the national standards institutes of 162 countries, one member per country, with a Central Secretariat in Geneva, Switzerland, that coordinates the system. ISO is a non-governmental organization that forms a bridge between the public and private sectors. An annual meeting is held of the Organization's General Assembly, the governing body. Many of the ISO's member institutes are part of the governmental structure of their countries, or are mandated by their government. Other members have their roots uniquely in the private sector, having been set up by national partnerships of industry associations. Therefore, ISO enables a consensus to be reached on solutions that meet both the requirements of business and the broader needs of society.[118]
Standards	<ul><li>ISO 19115-1: *Geographic information – Metadata – Part 1: Fundamentals*</li><li>ISO/IEC 20000-1: *Information technology – Service management – Part 1: Service management system requirements*</li><li>ISO/IEC 24760-1:2011: *Information technology – Security techniques – A framework for identity management – Part 1: Terminology and concepts*</li><li>ISO/IEC 24760-2:2015: *Information technology – Security techniques – A framework for identity management – Part 2: Reference architecture and requirements*</li><li>ISO/IEC 24760-3: *Information technology – Security techniques – A framework for identity management – Part 3: Practice* (under development)</li><li>The ISO 27000 family is a series of standards related to information security. Below is a selection of standards that can be applied to NG911 networks and operations. Please note that other standards in the ISO 27000 family also may be applicable to NG911 networks and operations.<ul><li>ISO/IEC 27000:2016: *Information technology – Security techniques – Information security management systems – Overview and vocabulary*</li><li>ISO/IEC 27001:2013: *Information technology – Security techniques – Information security management systems – Requirements* ISO/IEC 27002:2013: *Information technology – Security techniques – Code of practice for information security controls*</li></ul></li></ul>

---

[118] International Organization of Standards (ISO), *About ISO*. Available at: http://www.iso.org/iso/about.htm (last accessed February 26, 2016).

Standards (continued)	o ISO/IEC 27003:2010: *Information technology – Security techniques – Information security management system implementation guidance*
	o ISO/IEC 27004:2009: *Information technology – Security techniques – Information security management – Measurement*
	o ISO/IEC 27005:2011: *Information technology – Security techniques – Information security risk management*
	o ISO/IEC 27011:2008: *Information technology – Security techniques – Information security management guidelines for telecommunications organizations based on ISO/IEC 27002*
	o ISO/IEC 27031:2011: *Information technology – Security techniques – Guidelines for information and communication technology readiness for business continuity*
	o ISO/IEC 27033-1:2015: *Information technology – Security techniques – Network security – Part 1: Overview and concept*
	o IISO/IEC 27033-2:2012: *Information technology – Security techniques – Network security – Part 2: Guidelines for the design and implementation of network security*
	o ISO/IEC 27033-3:2010: *Information technology – Security techniques – Network security – Part 3: Reference Networking scenarios – Threats, design techniques and control issues*
	o ISO/IEC 27033-4:2014: *Information technology – Security techniques – Network security – Part 4: Securing communications between networks using security gateways*
	o ISO/IEC 27033-5:2013: *Information technology – Security techniques – Network security – Part 5: Securing communications across networks using Virtual Private Networks (VPNs)*
	o ISO/IEC 27035:2011: *Information technology – Security techniques – Information security incident management*
	o ISO/IEC 27037:2012: *Information technology – Security techniques – Guidelines for identification, collection, acquisition and preservation of digital evidence*

- ISO/IEC CD 29003: *Information technology – Security techniques – Identity proofing* (under development)
- ISO/IEC 29115:2013: *Information technology – Security techniques – Entity authentication assurance framework*
- ISO/IEC FDIS 29146: *Information technology – Security techniques – A framework for access management* (under development)

Website	http://www.iso.org/

International Telecommunication Union (ITU)

Name International Telecommunication Union (ITU) Standardization Section (ITU-T)

Type International Standards Organization

Summary Through its work on standardization, ITU develops technical standards (known as Recommendations) that facilitate the use of public telecommunication services and systems for communications during emergency, disaster relief, and mitigation operations. In such circumstances, technical features need to be in place to ensure that users who must communicate at a time of disaster have the communication channels they need, with appropriate security and with the best possible quality of service.[119]

Relevant Study Groups
- Study Group 2: Study Group 2 is responsible for the numbering standard ITU-T Recommendation, E.164, which has played a key role in shaping the telecommunications networks of today. E.164 provides the structure and functionality for telephone numbers; without it, individuals would not be able to communicate internationally. In recent years, Study Group 2 has worked on E.164 Number Mapping (ENUM), an IETF protocol for entering E.164 numbers into the Internet domain name system (DNS). A less well-known, but equally important product of Study Group 2 is E.212, which describes a system to identify mobile devices as they move from network to network. International mobile subscriber identity (IMSI) is a critical part of the modern mobile telecoms system, allowing a roaming mobile terminal to be identified in another network and subsequently for querying of the home network for subscription and billing information to take place.[120]
- Study Group 11: Study Group 11 is the "signaling" group within ITU-T; it produces ITU-T Recommendations that define how telephone calls and other calls such as data calls are handled in the network. Previously, this occurred primarily in the PSTN and the Integrated Services Digital Network (ISDN). Now, as operators look to align this circuit-switch-based environment with the rapidly emerging Internet technologies, Study Group 11's work is shifting toward IP-based networks or NGNs.[121]

---

[119] ITU, *Emergency Telecoms*. Available at: http://www.itu.int/en/ITU-T/emergencytelecoms/Pages/default.aspx (last accessed February 26, 2016).

[120] ITU, *Study Group 2 at a Glance*. Available at: http://www.itu.int/net/ITU-T/info/sg02.aspx (last accessed February 26, 2016).

[121] ITU, *Study Group 11 at a Glance*. Available at: http://www.itu.int/net/ITU-T/info/sg11.aspx (last accessed February 26, 2016).

Relevant Study Groups (continued)	• <u>Study Group 13</u>: Study Group 13 leads ITU's work on standards for NGNs. Broadly speaking, the term NGN refers to the move from circuit-switched to packet-based networks that many operators worldwide will undertake in the next few years. It will mean reduced costs for service providers who, in turn, will be able to offer a richer variety of services.[122]
Standards	• ITU-T P.800.2: *Mean opinion score interpretation and reporting* • ITU-T X.509: *Information technology – Open Systems Interconnection – The Directory: Public-key and attribute certificate frameworks* • ITU-T Y.1271: *Framework(s) on network requirements and capabilities to support emergency telecommunications over evolving circuit-switched and packet-switched networks* • ITU-T Y.2705: *Minimum security requirements for the interconnection of Emergency Telecommunications Service (ETS)*
Coordinated Activities	• IETF: In recent years, Study Group 2 has worked on ENUM, an IETF protocol for entering E.164 numbers into the Internet DNS.[123]
Website	http://www.itu.int/

---

[122] ITU, *Study Group 13 at a Glance*. Available at: http:/www.itu.int/net/ITU-T/info/sg13.aspx (last accessed February 26, 2016).

[123] ITU, ENUM. Available at: http://www.itu.int/osg/spu/enum/ (last accessed February 26, 2016.

ISACA®

Name	ISACA®
Type	Global Information Systems Security Organization
Summary	As an independent, non-profit, global association, ISACA engages in the development, adoption, and use of globally accepted, industry-leading knowledge and practices for information systems. Previously known as the Information Systems Audit and Control Association, ISACA now goes by its acronym only, to reflect the broad range of IT governance professionals it serves.[124]

ISACA helps global professionals lead, adapt, and assure trust in an evolving digital world by offering innovative and world-class knowledge, standards, networking, credentialing and career development. Established in 1969, ISACA is a global non-profit association of 140,000 professionals in 180 countries. ISACA also offers the Cybersecurity Nexus™ (CSX), a holistic cybersecurity resource, and COBIT®, a business framework to govern enterprise technology.[125]

COBIT 5 is the only business framework for the governance and management of enterprise IT.[126] It is the product of a global task force and development team from ISACA. COBIT 5 is generic and useful for enterprises of all sizes, whether commercial, not-for-profit or in public sector. COBIT 5 is used by those who have the primary responsibility for business processes and technology, depend on technology for relevant and reliable information, and provide quality, reliability and control of information and related technology. Key COBIT 5 users include enterprise executives and consultants in the following areas:
- Audit and Assurance
- Compliance
- IT Operations
- Governance
- Security and Risk Management

Mission	For professionals and organizations be the leading global provider of knowledge, certifications, community, advocacy and education on information systems assurance and security, enterprise governance of IT, and IT-related risk and compliance—contains in its few words a clear definition of ISACA's target audiences, its products and services, and its areas of professional expertise.

---

[124] ISACA, *About ISACA*. Available at: http://www.isaca.org/about-isaca/pages/default.aspx (last accessed February 26, 2016).

[125] ISACA, *ISACA Fact Sheet*. Available at: http://www.isaca.org/About-ISACA/Press-room/Documents/2015-ISACA-Fact-Sheet_pre_eng_1015.pdf (last accessed February 26, 2016).

[126] COBIT, *About COBIT 5*. Available at: https://cobitonline.isaca.org/about (last accessed February 26, 2016).

Relevant Committees	• ISACA coordinates and participates in numerous committees, working groups and advisory groups. Those who serve on ISACA's volunteer bodies provide ISACA with insights and expertise from around the world, facilitating the execution of ISACA's strategy while interacting and forming connections with peers worldwide. Additional information is available on ISACA's website and include:
	  o Government and Regulatory Agency (GRA)
	  o Certification

Relevant Projects	• <u>Voice-over Internet Protocol (VoIP) Audit/Assurance Program</u>: IT audit and assurance professionals are expected to use the program document to develop a customized program for the environment in which they are performing an assurance process related to information systems security. The document is to be used as a review tool and starting point and not intended to be a checklist or questionnaire. It is assumed that the IT audit and assurance professional has the necessary subject matter expertise required to conduct the work and is supervised by a professional with the Certified Information Systems Auditor (CISA) designation and/or necessary subject matter expertise to adequately review the work performed.
	
	A typical VoIP network comprises a complex series of cooperating protocols, networks (wireless and wired), servers, security architectures, special services (such as E-911), backup and recovery systems, and interfaces to the PSTN.
	
	During the audit planning process, the auditor must determine the scope of the audit. Depending on the specific implementation, this may include:
	  o Evaluation of governance, policies, and oversight relating to VoIP
	  o Data classification policies and management
	  o The appropriate VoIP business case, actual deployment or upgrade processes, strategy and implementation controls
	  o Technical architecture(s), including security systems, multiple platforms (different vendors which supply and/or support VoIP), interfaces with data networks, backup and recovery, data retention and destruction policy, and technology
	  o Assessments of IT infrastructure and personnel to support the VoIP architecture(s)
	  o Baseline configurations of deployed hardware and software
	  o Issues related to decentralized VoIP servers
	  o Issues related to failover clustering, where appropriate
	
	Security considerations for the PSTN (or dial-up) are outside the scope of this document.[127]

---

[127] *Voice-over Internet Protocol (VoIP) Audit/Assurance Program*. Available at: <u>http://www.isaca.org/Knowledge-Center/Research/ResearchDeliverables/Pages/Voice-over-Internet-Protocol-VoIP-Audit-Assurance-Program.aspx</u> (last accessed February 26, 2016).

Relevant Projects (continued)	• <u>COBIT 5 Assessment Programme</u>: The Programme is the basis for assessing an enterprise's processes for the governance and management of information technology and related services as described in COBIT 5. The Programme consists of the following segments:
	  ○ Process Assessment Model
	  ○ Self-Assessment Guide
	  ○ Assessor Guide
	
	It enables the evaluation of selected IT processes. The assessment results provide a determination of process capability and can be used for process improvement, delivering value to the business, measuring the achievement of current or projected business goals, benchmarking, consistent reporting and organizational compliance.
	
	The process capability is expressed in terms of attributes grouped into capability levels and the achievement of specific process attributes as defined in ISO/IEC 15504-2. Processes can be assessed individually or alternatively in logical groups. As such, scoping areas have been defined based on previously developed mappings, published by ISACA, which will allow for focused assessments. These scoping areas include:
	  ○ Capability of IT processes to support cloud services
	  ○ Capability of IT processes to support achievement of IT and business goals
	  ○ Capability of IT processes to support SOX compliance
	  ○ Capability of IT processes to support the enterprise governance of IT[128]
	
	• <u>CSX</u>: CSX is a security knowledge platform and professional program from ISACA. CSX is focused on helping shape the future of cybersecurity through cutting-edge thought leadership, as well as training and certification programs for the professionals who are leading it there. Building on the strength of ISACA's globally-recognized expertise, it gives cybersecurity professionals a smarter way to keep organizations and their information more secure. With CSX, business leaders and cyber professionals can obtain the knowledge, tools, guidance and connections to be at the forefront of a vital and rapidly changing industry.[129]

Standards	• ISACA engages in the development, adoption, and use of globally accepted, industry-leading knowledge and practices for information systems, but is not an SDO.

---

[128] *COBIT 5 Assessment Programme*. Available at: http://www.isaca.org/COBIT/Pages/Product-Family.aspx#process (last accessed February 26, 2016).

[129] *Cybersecurity Nexus™ (CSX)*. Available at: http://www.isaca.org/cyber/Pages/default.aspx (last accessed February 26, 2016).

Coordinated Activities	• ISACA is a global non-profit association of professionals in 180 countries that collaborate to help global professionals lead, adapt and assure trust in an evolving digital world by offering innovative and world-class knowledge, standards, networking, credentialing and career development.
Websites	https://www.isaca.org/ https://cobitonline.isaca.org/

National Emergency Number Association (NENA)

Name National Emergency Number Association (NENA)

Type National Standards Organization (ANSI-accredited)

Summary NENA serves its members and the greater public safety community as the only professional organization solely focused on 9-1-1 policy, technology, operations, and education issues. With more than 7,000 members in 48 chapters across the U.S. and around the globe, NENA promotes implementation and awareness of 9-1-1, as well as international three-digit emergency communications systems. NENA is an ANSI-accredited standards developer.

NENA works with 9-1-1 professionals nationwide, public policy leaders, emergency services and telecommunications industry partners, like-minded public safety associations, and other stakeholder groups to develop and carry out critical programs and initiatives; to facilitate the creation of an IP-based NG9-1-1 system; and to establish industry-leading standards, training, and certifications. Through the association's efforts to provide effective and efficient public safety solutions, NENA strives to protect human life, preserve property, and maintain the security of our communities.

NENA began work on what is now termed NG9-1-1 in 2000 with discussion and then production of the NENA Future Path Plan for a technologically updated and more feature-rich replacement for E9-1-1. In 2003, NENA established a committee to develop the technical nature and architecture of NG9-1-1, recognizing that this would also require various other work efforts over time to define databases management, system operations and administration, and PSAP operations requirements and standards, as well as transition plans. The NENA NG9-1-1 Project was formed to tie all aspects together.

Relevant Committees

- NENA NG9-1-1 Project encompasses and coordinates many actions aimed to accomplish the capabilities for IP-based NG9-1-1:
  - Core systems and technical development
  - PSAP operations
  - NG9-1-1 system operations
  - Interconnection and security
  - Policy change needs and methods development
  - Transition planning
  - Public education and PSAP training
  - Interoperability testing (Industry Collaboration Events [ICE])

There are also plans to conduct a distributed Pilot Testing process to result in national testing recommendations.

Standards

- Data and Network Standards
  - NENA 02-010 v9: *NENA Standard Data Formats For 9-1-1 Data Exchange & GIS Mapping*
  - NENA 02-011 v7.1: *NENA Data Standards For Local Exchange Carriers, ALI Service Providers & 9-1-1 Jurisdictions*
  - NENA 02-014 v1: *NENA GIS Data Collection and Maintenance Standards*
  - NENA 02-015 v1: *NENA Technical Standard for Reporting and Resolving ANI/ALI Discrepancies and No Records Found for Wireline, Wireless and VoIP Technologies*
  - NENA 03-509 v1: *Femtocell and UMA TID*
  - NENA 04-005 v1: *NENA ALI Query Service Standard*
  - NENA 06-750 v3: *NENA Technical Requirements Document On Model Legislation E9-1-1 for Multi-Line Telephone Systems*
  - NENA 08-001 v2: *NENA Interim VoIP Architecture for Enhanced 9-1-1 Services (i2)*
  - NENA 08-503 v1: *VoIP Characteristics Technical Information Document*
  - NENA 08-505 v1: *NENA Recommended Method(s) for Location Determination to Support IP-Based Emergency Services*
  - NENA 08-752 v1: *NENA Technical Requirements Document For Location Information to Support IP-Based Emergency Services*
  - NENA 71-001 v1: *NENA Standard For NG9-1-1 Additional Data* (update in progress)
  - NENA 71-501: *Synchronizing GIS System Databases with MSAG & ALI*
  - NENA-STA-004.1.1-2014: *NENA Next Generation 9-1-1 (NG9-1-1) Civic Location Data Exchange Format (CLDXF) Standard*
  - NENA-STA-006.1-201X: *GIS Data Model for NG9 1-1* (in progress)
  - NENA-INF-009.1-2014: *Requirements for a National Forest Guide Information Document*

Standards (continued)

- o NENA/APCO-INF-005: *NENA/APCO Emergency Incident Data Document (EIDD) Information Document*
- o NENA 70-Draft: *Provisioning and Maintenance of GIS data to ECRF/LVF* (in progress)
- o NENA TBD: *Discrepancy, Performance and Audits for NG9-1-1* (in progress)
- o NENA-INF-014.1-2015: *NENA Information Document for Development of Site/Structure Address Point GIS Data for 9-1-1*
- Policy Routing Standards
  - o NENA 71-502 v1: *An Overview of Policy Rules for Call Routing and Handling in NG9-1-1*
  - o NENA STA-003.1.1-2014: *NENA Standard for NG9-1-1 Policy Routing Rules*
  - o NENA-INF-011.1-2014: *NENA NG9-1-1 Policy Routing Rules Operations Guide*
- Security Standards
  - o NENA 75-001: *NENA Security for Next-Generation 9-1-1 Standard (NG-SEC)*
  - o NENA 75-502 v1: *Next Generation 9-1-1 Security (NG-SEC) Audit Checklist*
- NG9-1-1 Architecture Standards
  - o NENA 08-002 v1: *NENA Functional and Interface Standards for Next Generation 9-1-1 Version 1.0 (i3)*
  - o NENA 08-003 v1: *Detailed Functional and Interface Specification for the NENA i3 Solution – Stage 3* (update in progress; to be renumbered as NENA-STA-010)
  - o NENA 08-501 v1: *NENA Technical Information Document on the Network Interface to IP Capable PSAP*
  - o NENA-08-506 v1: *NENA Emergency Services IP Network Design for NG9-1-1 (NID)*
  - o NENA 08-751 v1: *NENA i3 Technical Requirements Document*
  - o NENA 53-507: *NENA Virtual PSAP Management Operations Information Document (OID)*
  - o NENA 73-501 v1: *Use Cases & Suggested Requirements for Non-Voice-Centric (NVC) Emergency Services*
  - o NENA-INF-003.1-2013: *NENA Potential Points of Demarcation in NG9-1-1 Networks Information Document*
  - o NENA-INF-TBD: *Non-Mobile Wireless and Broadband Connectivity*
  - o NENA-INF-TBD: *NENA Classes of Service*
  - o NENA/APCO-REQ-001.1.1-2016: *NENA/APCO Next Generation 9-1-1 Public Safety Answering Point Requirements*
- PSAP Operations, Training and Public Education Standards
  - o NENA 54-750 v1: *NENA/APCO Human Machine Interface & PSAP Display Requirements (ORD)*
  - o NENA 57-750 v1: *NG9-1-1 System and PSAP Operational Features and Capabilities Requirements*

Standards (continued)	o NENA 04-004: *NENA Recommended Generic Standards for E9-1-1 PSAP Intelligent Workstations*
	o NENA-INF-007.1-2013: *NENA Information Document for Handling Text-to-9-1-1 in the PSAP*
	o NENA-INF-012.2-2015: *NENA Inter-Agency Agreements Model Recommendations Information Document*
	o NENA-REF-002.2-2014: *PSAP Interim Text-to-9-1-1 Support Documents*
	o NENA-REF-003.1-2015: *NENA Text-to-9-1-1 Public Education*
	o *SMS Text-to-9-1-1 Resources for PSAPs and 9-1-1 Authorities*
	o *9-1-1 Authorities Guide to NG9-1-1*
	o *Recommended NG9-1-1 Public Education Plan for Elected Officials and Decision Makers*
	• (Management of) NG9-1-1 System Operations
	    o NENA-STA-008.2-2014: *NENA Registry System Standard*
	    o NENA-INF-TBD: *Monitoring and Managing NG9-1-1*
	• Transition Standards
	    o NENA-INF-008.2-2013: *NENA NG9-1-1 Transition Plan Considerations Information Document*
	    o NENA-INF-006.1-2014: *NENA NG9-1-1 Planning Guidelines Information Document*
	    o *Next Generation 9-1-1 Transition Policy Implementation Handbook*
	• Reference Standards
	    o NENA-ADM-000.18-2014: *NENA Master Glossary of 9-1-1 Terminology*

Coordinated Activities	• USDOT NG911 Initiative • Integrated Justice Information Systems (IJIS) • Next Generation Partner Program (NGPP) coordinates with various industry vendors and public safety groups • NG9-1-1 ICE coordinates with industry vendors on interoperability and standards compliance • ATIS ESIF regarding emergency services interconnection issues • N11 consortium for coordinating interactions between NG911 and N11 services • Coalition of Geospatial Organizations (COGO) • Urban and Regional Information Systems Association (URISA) • National Center for Missing and Exploited Children (NCMEC) • FCC CSRIC / TFOPA • ANSI: NENA is an ANSI-accredited SDO. • Implementation and Coordination Office (ICO) 911 Resource Center
Effects on NG911	• Defines ESInet (transport and connectivity) requirements and characteristics, beyond generic IP networking standards

Effects on NG911 (continued)

- Defines NG9-1-1 IP function and interface standards for NG9-1-1 core architecture
- Defines NG9-1-1 databases used to control call-routing processes
- Supports location requirements and standards
- Defines NG9-1-1 interface options for originating service provider entry to the system
- Defines emergency entity functionality in coordination with NG9-1-1 system functions
- Defines PSAP functional entity downstream interfaces
- Defines mechanisms for acquisition of additional data from beyond the NG9-1-1 system
- Addresses PSAP operations

Website

http://www.nena.org/
http://www.nena.org/?page=Standards

National Fire Protection Association (NFPA)

Name   National Fire Protection Association (NFPA)

Type   National Standards Organization (ANSI-accredited)

Summary   NFPA is the world's leading advocate of fire prevention and an authoritative source on public safety. It develops, publishes, and disseminates more than 300 consensus codes and standards intended to minimize the possibility and effects of fire and other risks.[130]

Standards
- NFPA 70: *National Electrical Code® (NEC)*
- NFPA 72: *National Fire Alarm and Signaling Code*
- NFPA 76: *Standard for the Fire Protection of Telecommunications Facilities*
- NFPA 1061: *Professional Qualifications for Public Safety Telecommunications Personnel*
- NFPA 1201: *Standard for Providing Fire and Emergency Services to the Public*
- NFPA 1221: *Standard for the Installation, Maintenance, and Use of Emergency Services Communications Systems*
- NFPA 1600: *Standard on Disaster/Emergency Management and Business Continuity/Continuity of Operations Programs*

Coordinated Activities
- ANSI: NFPA is an ANSI-accredited SDO[131]

Website   http://www.nfpa.org/

---

[130] NFPA, *About NFPA.* Available at: http://www.nfpa.org/about-nfpa (last accessed February 26, 2016).
[131] NFPA, *Overview.* Available at: http://www.nfpa.org/about-nfpa/nfpa-overview (last accessed February 26, 2016).

National Information Exchange Model (NIEM)

Name	National Information Exchange Model (NIEM)
Type	Government Project
Summary	NIEM was formed in 2005 and is a partnership between DOJ, DOC, and the Department of Health and Human Services (DHHS). NIEM connects communities of people who share a common need to exchange information in order to advance their mission.[132] Organizations who use NIEM can benefit from enhanced mission capabilities, reduced costs, and offer community support.[133] All 50 states and the majority of federal agencies are using (at varying levels of maturity) or considering using NIEM.
Relevant Programs and Projects	• NIEM Program Management Office (PMO): The PMO executes the vision of NIEM established by the Executive Steering Council (ESC), while managing the day-to-day operations of NIEM. The office encourages the adoption and use of NIEM and oversees all working group and committee activities, regularly coordinating with communities of interest, principal stakeholders, and other information-sharing initiatives to promote collaboration and interest in NIEM priorities.[134] • NIEM Business Architecture Committee (NBAC): NBAC operates in coordination with NTAC and PMO to meet the goals of the project as a whole. NBAC's mission is to set the business architecture and requirements of NIEM, manage NIEM core and facilitate the processes for the regulation and support of NIEM domains.[135] • NIEM Technical Architecture Committee (NTAC): NTAC operates in coordination with NBAC and PMO to meet the goals of the project as a whole. NTAC's mission is to define and support the technical architecture that governs NIEM.[136]

---

[132] National Information Exchange Model, *About NIEM*. Available at: http://www.niem.gov (last accessed February 26, 2016).

[133] National Information Exchange Model, *Roadmap to Adoption*. Available at: https://www.niem.gov/aboutniem/roadmap/Pages/niem-engagement.aspx (last accessed February 26, 2016).

[134] National Information Exchange Model, *NIEM Program Management Office*. Available at: https://www.niem.gov/meet-us/Pages/pmo.aspx (last accessed February 26, 2016).

[135] National Information Exchange Model, *NIEM Business Architecture Committee*. Available at: https://www.niem.gov/meet-us/nbac (last accessed February 26, 2016).

[136] National Information Exchange Model, *NIEM Technical Architecture Committee*. Available at: https://www.niem.gov/meet-us/ntac (last accessed February 26, 2016).

Relevant Programs and Projects (continued)	• APCO/NENA Emergency Incident Data Document (EIDD) Working Group: The goal of the APCO/NENA EIDD Working Group is to initiate the process of creating a NIEM conformant, ANS that will be used to share emergency incident information between and among authorized entities and systems.[137]
Standards	• NIEM *version 3.1*
Coordinated Activities	• The Geospatial Enhancement for NIEM (Geo4NIEM) initiative is a collaboration between NIEM PMO, the Open Geospatial Consortium (OGC), DHS and the Program Manager for the Information Sharing Environment to enhance the capabilities of NIEM's geospatial exchange.[138]
	• The EIDD provides formats for sharing emergency incident information in the next generation environment. The content of the EIDD provides the audience with the recommended list of data components, their relationships to each other, the data elements contained within each data component, and where applicable the registries that control the available values for appropriate data elements. This list of data components and data elements, along with their identified attributes and allowable values, when finalized, will become the basis for the NIEM-conformant Information Exchange Package Document ANS, along with its eXtensible Markup Language (XML) schema and associated artifacts.
Effects on NG911	• Develops standards related to handling emergency data sets, specifically pertaining to interoperability for data sharing.
Website	http://niem.gov

---

[137] National Emergency Number Association, *APCO/NENA EIDD Working Group*. Available at: https://www.nena.org/?EIDD (last accessed February 26, 2016).
[138] Geospatial for NIEM, National Information Exchange Model, Available at: https://www.niem.gov/technical/Pages/Geo4NIEM.aspx (last accessed February 26, 2016).

North American Electric Reliability Corporation (NERC)

Name	North American Electric Reliability Corporation (NERC)
Type	Professional Organization
Summary	NERC is a not-for-profit international regulatory authority whose mission is to assure the reliability of the bulk power system in North America. NERC develops and enforces Reliability Standards; annually assesses seasonal and long-term reliability; monitors the bulk power system through system awareness; and educates, trains, and certifies industry personnel. NERC's area of responsibility spans the continental U.S., Canada, and the northern portion of Baja California, Mexico. NERC is the electric reliability organization for North America, subject to oversight by the Federal Energy Regulatory Commission and governmental authorities in Canada. NERC's jurisdiction includes users, owners, and operators of the bulk power system, which serves more than 334 million people.[139]
	NERC has nine standards that are subject to enforcement by NERC. There are additionally 17 standards subject to future enforcement and 3 pending regulatory filings.[140]
Relevant Committees	• Standards Committee • Critical Infrastructure Protection Committee
Standards	• CIP-002-5.1: *Cyber Security — BES Cyber System Categorization* • CIP-003-5: *Cyber Security — Security Management Controls* • CIP-004-5.1: *Cyber Security — Personnel & Training* • CIP-005-5: *Cyber Security — Electronic Security Perimeter(s)* • CIP-006-5: *Cyber Security — Physical Security of BES Cyber Systems* • CIP-007-5: *Cyber Security — System Security Management* • CIP-008-5: *Cyber Security — Incident Reporting and Response Planning* • CIP-009-5: *Cyber Security — Recovery Plans for BES Cyber Systems* • CIP-010-1: *Cyber Security — Configuration Change Management and Vulnerability Assessments*
Effects on NG911	• These standards apply to the electrical critical infrastructure and will not have direct impact on NG911. This level of cyber security for critical infrastructure is in line with what is needed for NG911.
Website	http://www.nerc.com/

---

[139] North American Electric Reliability Corporation website. At http://www.nerc.com/Pages/default.aspx (last accessed February 26, 2016).

[140] North American Electric Reliability Corporation website. At http://www.nerc.com/pa/Stand/Pages/CIPStandards.aspx (last accessed February 26, 2016).

Object Management Group® (OMG®)

Name	Object Management Group® (OMG®)
Type	Not-for-profit technology standards consortium
Summary	OMG is an international, open membership, not-for-profit technology standards consortium, founded in 1989. OMG standards are driven by vendors, end-users, academic institutions and government agencies. OMG Task Forces develop enterprise integration standards for a wide range of technologies and an even wider range of industries. OMG's modeling standards, including the Unified Modeling Language® (UML®) and Model Driven Architecture® (MDA®), enable powerful visual design, execution and maintenance of software and other processes. OMG also hosts organizations such as the user-driven information-sharing Cloud Standards Customer Council™ and the IT industry software quality standardization group, the Consortium for IT Software Quality™. OMG also managed the Industrial Internet Consortium, the public-private partnership that was formed in 2014 with AT&T, Cisco, GE, IBM, and Intel to forward the development, adoption, and innovation of the Industrial IoT. [141]
Mission Statement	OMG's mission is to develop technology standards that provide real-world value for thousands of vertical industries. OMG is dedicated to bringing together its international membership of end-users, vendors, government agencies, universities and research institutions to develop and revise these standards as technologies change throughout the years.[142]
Relevant Documents	• Cyber Security Protection for Front Line Real-Time Systems RFI: This Request for Information (RFI) solicits information about requirements, standards, products, and work in progress as well as prospective work related to Security Services along with digital (non-XML) Security Data Tagging and other Security-related Services. OMG and, specifically, the coordination task force C4I DTF, and related task forces such as MARS PTF, System Assurance (SysA), Government, DDS PSIG, SysML PSIG and other groups within OMG, will use this information to begin the process for OMG-compliant models and interface standards to be used in platforms. [143]

---

[141] OMG Website, *About OMG*. Available at: http://www.omg.org/gettingstarted/gettingstartedindex.htm (last accessed February 26, 2016)

[142] OMG Website, *About OMG*. Available at: http://www.omg.org/gettingstarted/gettingstartedindex.htm (last accessed February 26, 2016)

[143] OMG Website, *Current OMG Technology Adoption Processes Under Way*. Available at: http://www.omg.org/public_schedule/ (last accessed February 26, 2016).

Relevant Documents (continued)	• UML Operational Threat & Risk Model RFP: Multiple communities have developed data and exchange schema and interfaces for sharing information about threats, risks and incidents that impact important government, commercial, and personal assets and privacy. This RFP calls for a conceptual model for operational threats and risks that unifies the semantics of and can provide a bridge across multiple threat and risk schema and interfaces. The conceptual model will be informed by high-level concepts as defined by the Cyber domain, existing NIEM domains and other applicable domains, but is not specific to those domains. This will enable combined Cyber, physical, criminal and natural threats and risks to be federated, understood and responded to effectively.[144]
Coordinated Activities	OMG maintains active relationships with many other standards bodies and consortia such as:[145] • 3GPP • IJIS Institute • ISO • ITU-T Standardization Sector • OASIS • Open GIS Consortium
Effects on NG911	• Current work on cybersecurity and risk may benefit the NG911 environment.
Website	http://www.omg.org

---

[144] OMG Website, *Current OMG Technology Adoption Processes Under Way*. Available at: http://www.omg.org/public_schedule/ (last accessed February 26, 2016).

[145] OMG Website, Liaison AB Subcommittee, Available at: http://www.omg.org/news/about/liaison.htm (last accessed February 26, 2016).

Organization for the Advancement of Structured Information Standards (OASIS)

Name Organization for the Advancement of Structured Information Standards (OASIS)

Type Standards Setting Organization (Community)

Summary OASIS is a not-for-profit consortium that drives the development, convergence, and adoption of open standards for the global information society.[146]

Relevant Committees
- OASIS Emergency Management Technical Committee (EM-TC): The mission of the EM-TC is to create incident- and emergency-related standards for data interoperability. The EM-TC welcomes participation from members of the emergency management community, developers and implementers, and members of the public concerned with disaster management and response.[147]

Standards
- OASIS CAP v1.2: *Common Alerting Protocol*
- OASIS EDXL-DE v1.0: *Emergency Data Exchange Language (EDXL) Distribution Element, v 1.0*
- OASIS EDXL-HAVE: *Emergency Data Exchange Language (EDXL) Hospital AVailability Exchange Version 1.0*
- OASIS EDXL-RM: *Emergency Data Exchange Language Resource Messaging (EDXL-RM) 1.0*
- OASIS EDXL-SitRep 1.0: *Emergency Data Exchange Language Situation Reporting (EDXL-SitRep) Version 1.0*
- OASIS EDXL-TEC: *Emergency Data Exchange Language (EDXL) Tracking of Emergency Clients (TEC) Client Registry Exchange Version 1.0*
- OASIS EDXL-TEP v1.1: *Emergency Data Exchange Language (EDXL) Tracking of Emergency Patients (TEP) Version 1.1* (in progress)

Effects on NG911
- Develops standards related to handling emergency data sets.

Website http://www.oasis-open.org/

---

[146] OASIS, *About OASIS*. Available at: http://www.oasis-open.org/org (last accessed February 26, 2016).
[147] OASIS, *OASIS Emergency Management TC*. Available at: http://www.oasis-open.org/committees/tc_home.php?wg_abbrev=emergency (last accessed February 26, 2016).

Open Geospatial Consortium (OGC®)

Name	Open Geospatial Consortium (OGC®)
Type	Standards Setting Organization (Community)
Summary	OGC is an international industry consortium of over 500 companies, government agencies, and universities participating in a consensus process to develop publicly available interface standards. OGC® Standards support interoperable solutions that "geo-enable" the Web, wireless and location-based services, and mainstream IT. The standards empower technology developers to make complex spatial information and services accessible and useful with all kinds of applications.[148]
Mission	OGC's mission is to advance the development and use of international standards and supporting services that promote geospatial interoperability. To accomplish this mission, OGC serves as the global forum for the collaboration of geospatial data/solution providers and users.

Standards

- OGC 04-094: *Web Feature Service Implementation Standard*
- OGC 06-042: *OpenGIS® Web Map Server Implementation Specification*
- OGC 07-006r1: *OpenGIS® Catalogue Services Specification*
- OGC 07-074: *OpenGIS® Location Service (OpenLS): Core Services*
- OGC 09-025r2: *OGC® Web Feature Service 2.0 Interface Standard*
- OGC 09-083r3: *GeoAPI 3.0 Implementation Standard*
- OGC 10-129r: *OGC® Geography Markup Language (GML) – Extended schemas and encoding rules*
- OGC 11-030R1: *OGC® Open GeoSMS Standard – Core*
- OGC 12-019: *OGC City Geography Markup Language (CityGML) Encoding Standard*
- OGC KML 2.3: *OGC KML 2.3*

Alliance Partners/ Coordinated Activities[149]

- IEEE
- IETF
- ISO
- OASIS
- OMA

---

[148]OGC, *OGC Vision, Mission, & Goals.* Available at: http://www.opengeospatial.org/ogc/vision (last accessed February 26, 2016).

[149] OGC, *OGC Alliance Partners.* Available at: http:// http://www.opengeospatial.org/ogc/alliancepartners (last accessed February 26, 2016).

Effects on NG911	• Supports geospatial data standards for data sharing, implementation and interoperability.
Website	http://www.opengeospatial.org/

Open Mobile Alliance (OMA)

Name	Open Mobile Alliance (OMA)
Type	International Standards Organization
Summary	OMA is the focal point for the development of mobile service enabler specifications, which support the creation of interoperable end-to-end mobile services. OMA drives service enabler architectures and open enabler interfaces that are independent of the underlying wireless networks and platforms. OMA creates interoperable mobile data service enablers that work across devices, service providers, operators, networks, and geographies. Toward that end, OMA will develop test specifications, encourage third-party tool development, and conduct test activities that allow vendors to test their implementations.[150]

Goals

- Deliver high quality, open technical specifications based upon market requirements that drive modularity, extensibility, and consistency amongst enablers to reduce industry implementation efforts.
- Ensure OMA service enabler specifications provide interoperability across different devices, geographies, service providers, operators, and networks; facilitate interoperability of the resulting product implementations.
- Be the catalyst for the consolidation of standards activity within the mobile data service industry; work in conjunction with other existing standards organizations and industry fora to improve interoperability and decrease operational costs for all involved.
- Provide value and benefits to members in OMA from all parts of the value chain including content and service providers, information technology providers, mobile operators and wireless vendors, such that they elect to actively participate in the organization.[151]

Relevant Working Groups

- Location Working Group: The Location Working Group develops specifications to ensure interoperability of mobile location services on an end-to-end basis, as well as to provide technical expertise and consultancy on location services for other groups within OMA.[152]

---

[150] OMA, *About OMA*. Available at: http://www.openmobilealliance.org/AboutOMA/Default.aspx (last accessed: February 26, 2016).

[151] OMA, *About OMA*. Available at: http://www.openmobilealliance.org/AboutOMA/Default.aspx (last accessed: February 26, 2016).

[152] OMA, *Location Working Group*. Available at: http://technical.openmobilealliance.org/Technical/technical-information/working-groups-and-committees/location-working-group (last accessed February 26, 2016).

Relevant Working Groups (continued)	• <u>Device Management Working Group</u>: The goal of the Device Management Working Group is to specify protocols and mechanisms that achieve management of mobile devices, including the necessary configuration to access services and management of the software on mobile devices.[153]
Standards	• OMA-ERP-SUPL-V3_O_2-20110920-C: *OMA Secure User Plane Location V3.0* • OMA-ERELD-LPPe-V2_0-20141202-C: *OMA LPP Extensions (LPPe) v2.0* • OMA-ERP-MLP-V3_1-20110920-A: *OMA Mobile Location Protocol V3.1* • OMA-ERELD-LOCSIP-V1_0-20120117-A: *OMA Location in SIP/IP Core V1.0* • OMA SEC_CF 1.1: *OMA Application Layer Security Common Functions V1.1*
Coordinated Activities	• 3GPP: OMA and 3GPP work to update on a regular basis the list of dependencies between each organization's specifications and work in progress • 3GPP2: OMA and 3GPP2 work to update on a regular basis the list of dependencies between each organization's specifications and work in progress • IETF: OMA and IETF work to update on a regular basis the list of dependencies between each organization's specifications and work in progress[154]
Effects on NG911	• Develops standards that enable text and multimedia transmission from the caller to the NG911 system (transport of data). • Supports location requirements and/or specifies standards.
Website	http://www.openmobilealliance.org/

---

[153] OMA, *Device Management Working Group*. Available at: http://technical.openmobilealliance.org/Technical/technical-information/working-groups-and-committees/device-management-working-group (last accessed February 26, 2016).
[154] OMA, *Collaborating with OMA*. Available at: http://openmobilealliance.org/about-oma/collaborating-with-oma/ (last accessed February 26, 2016).

Society of Cable Telecommunications Engineers (SCTE)

Name   Society of Cable Telecommunications Engineers (SCTE)

Type   Standards Setting Organization—Industry (Cable Telecommunications)

Summary   SCTE is a non-profit professional association that provides technical leadership for the telecommunications industry and serves its members through professional development, standards, certification, and information.[155]

Mission   SCTE's mission is to provide technical leadership for the telecommunications industry and serve its members through excellence in professional development, standards, certification, and information.

Standards
- ANSI/SCTE 24-1 2009: *IPCablecom 1.0 Part 1: Architecture Framework for the Delivery of Time Critical Services Over Cable Television Networks Using Cable Modems*
- ANSI/SCTE 24-2 2009: *IPCablecom 1.0 Part 2: Audio Codec Requirements for the Provision of Bi-directional Audio Service Over Cable Television Networks Using Cable Modems*
- ANSI/SCTE 24-3 2009: *IPCablecom Part 3: Network Signaling Protocol for the Delivery of Time-Critical Services over Cable Television Using Data Modems*
- ANSI/SCTE 24-4 2009: *IPCablecom Part 4: Dynamic Quality of Service for the Provision of Real-time Services over Cable Television Networks using Cable Modems*
- ANSI/SCTE 24-21 2012: *BV16 Speech Codec Specification for Voice over IP Applications in Cable Telephony*
- ANSI/SCTE 24-22 2013: *iLBCv2.0 Speech Codec Specification for Voice over IP Applications in Cable Telephony*
- ANSI/SCTE 24-23 2012: *BV32 Speech Codec Specification for Voice over IP Applications in Cable Telephony*
- ANSI/SCTE 165-12 2009: *IPCablecom 1.5 Part 12: PSTN Gateway Call Signaling Protocol*
- CEA/SCTE J-STD-42-B: *Emergency Alert Messaging for Cable*

---

[155] Society of Cable Telecommunications Engineers, *About SCTE*. Available at: http://www.scte.org/SCTE/About (last accessed February 26, 2016).

Coordinated Activities

- ANSI: The SCTE Standards Program provides an ANSI-accredited forum for development of technical specifications supporting the cable telecommunications industry.[156]

Website http://www.scte.org/

---

[156] Society of Cable Telecommunications Engineers, *SCTE Standards Program.* Available at: http://www.scte.org/SCTE/Standards (last accessed February 26, 2016).

Standards Coordinating Council (SCC)

Name	Standards Coordinating Council (SCC)
Summary	SCC grew out of a need for a high-level view of the information sharing and safeguarding standards landscape. As information sharing standards grow and develop, there needed to be a body to oversee this development and provide guidance to SDOs on issues related to these standards and how they fit in the context of the overall landscape of information sharing and safeguarding initiatives.
	Developed in accordance with the Intelligence Reform and Terrorism Prevention Act of 2004, SCC serves as this high-level oversight, and the SCC provides advice and counsel to the standards development stakeholder community on matters related to information sharing and safeguarding standards.[157]
Goals	In addition to providing advice and counsel, the SCC is intended to advance responsible information sharing and safeguarding and information sharing standards. To do this, the SCC will:

- Identify high-priority standards activities that can be coordinated across SDOs for greater return on resources.
- Communicate stakeholder requirements to identify opportunities to develop or integrate technical and functional roadmaps.
- Coordinate governance processes across SDOs to streamline standards development activities and to enhance communication, collaboration, and consensus between standards partners.
- Coordinate outreach and training opportunities to reach a broader constituency.
- Coordinate private sector standards activities with federal governance bodies such as the Federal CIO Council.

Relevant Committees	• Project Interoperability: Project Interoperability is a start-up guide for information interoperability. Information interoperability is the ability to transfer and use information in a consistent, efficient way across multiple organizations and IT systems to accomplish operational missions. From a technical perspective, interoperability is developed through the consistent application of design principles and design standards to address a specific mission problem.[158]

---

[157] Standards Coordination Council Website, *About the Standards Coordinating Council*. Available at: http://www.standardscoordination.org/about (last accessed February 26, 2016).
[158] Standards Coordination Council Website, *Project Interoperability*. Available at: http://www.standardscoordination.org/project_interoperability (last accessed February 26, 2016)

Relevant Committees (continued)

- <u>Incident Management Information Sharing (IMIS)</u>: Current systems that are implemented to support the emergency management community are typically developed using proprietary data models that may inhibit information sharing. The IMIS Framework recommends an architecture where standard information encodings and standard data services are implemented in the IMIS-compliant systems to support the necessary data exchanges and improve overall interoperability between systems holding information and systems supporting consumers of the information.[159]

- <u>Cross-border Trusted Information Sharing</u>: This effort plans to address cross-border information sharing between Canada and U.S. emergency management mission partners. Standardized information sharing will provide the mission partners across the borders to maintain and reach back into the situation awareness information, and where conditions match, a standardized message generated and sent based on pre-defined policies (rules). The goal of the project is to validate and enhance the IEF architecture.[160]

- <u>OGC Geo4NIEM Testbed</u>: Testing NIEM IC IEPs containing geospatial data or GML feature representations leveraging NIEM components to:
  - Validate and provide recommendations to enhance NIEM 3.0 architecture related to the Intelligence Community data encoding specifications (i.e., ISM, NTK, and TDF) aligned to OGC Testbed 9.
  - Provide recommendations to enable full-round tripping from NIEM information exchange packages to GML features and back to provide a comprehensive view of NIEM and GML capabilities and to document NIEM architectural gaps.
  - Test and demonstrate 1) use of NIEM 3.0 tagging related to IC data encoding specifications and 2) round tripping of NIEM information exchange packages to GML features and back.
  - Test and demonstrate use of an application programming interface (API) for operating on GML feature representations leveraging NIEM components; features may be searched, retrieved, inserted, updated, and deleted.[161]

---

[159] Standards Coordination Council Website, *SCC Initiatives*, Available at: http://www.standardscoordination.org/scc_initiatives (last accessed February 26, 2016).
[160] Standards Coordination Council Website, *SCC Initiatives*, Available at: http://www.standardscoordination.org/scc_initiatives (last accessed February 26, 2016).
[161] Standards Coordination Council Website, *SCC Initiatives*, Available at: http://www.standardscoordination.org/scc_initiatives (last accessed February 26, 2016).

Standards

- Information Sharing Environment (ISE) Information Interoperability Framework (I²F): ISE I²F is a national architecture framework designed to support information sharing for the public safety and national security missions across all levels of government – federal, state, local, tribal, and territorial. The content of Project Interoperability comes directly from the I²F.
- *Information Sharing and Safeguarding (IS&S) Playbook:* This Playbook is intended to help users in their quest to create or enhance an effective and efficient Information Sharing and Safeguarding environment, and can be used at any point in the environment's lifecycle.

Effects on NG911

- Impacts the sharing of information with other public safety and homeland security entities.

Website

http://www.standardscoordination.org

Telcordia

Name   Telcordia (now part of Ericsson)

Type   General requirements documents and reports for the Telecommunications Industry

Summary   Telcordia is a for-profit subsidiary of Ericsson, Inc. Telcordia provides vendor-neutral technical documentation and roadmaps to implementation of new technologies from the central office perspective. [162]

Mission   The Telcordia Information SuperStore features information products that are widely utilized, referenced, and accepted worldwide. Many of the Generic Requirements documents and special reports on telecommunications equipment, systems, and services are developed with industry participation, making them timely, high-quality, vendor-neutral technical specifications that are valuable to suppliers and service providers.

Standards
- SR-4163: *E9-1-1 Service Description*
- GR-1298: *ANGR: Switching Systems*
- FR-EMERG-SVCS-ARCH-01: *Family of Requirements for Emergency Services Network Architecture*
  - GR-3165: *Emergency Services Border Control Function (BCF) Generic Requirements*
  - GR-3166: *Legacy Public Safety Answering Point (PSAP) Gateway Generic Requirements*
  - GR-3170: *Legacy Selective Router (SR) Gateway Generic Requirements*
- FR-NEBS-EQUIP-PROTECT-01: *NEBS™ Family of Requirements for Protecting Network Equipment*
  - SR-3580: *NEBS™ Criteria Levels*

Coordinated Activities
- Coordination with telecommunications equipment manufacturers and the carrier industry to establish common interfaces and system requirements to ensure resiliency and interoperability. [163]

Website   http://telecom-info.telcordia.com/site-cgi/ido/docs2.pl?ID=056048165&page=home

---

[162] Telcordia, *Information SuperStore*. Available at: http://telecom-info.telcordia.com/site-cgi/ido/docs2.pl?ID=124445520&page=home (last accessed February 26, 2016).

[163] Telcordia, *NEBS Documents and Technical Services*. Available at: http://telecom-info.telcordia.com/site-cgi/ido/docs2.pl?ID=135181799&page=nebs (last accessed February 26, 2016).

Telecommunications Industry Association (TIA)

Name	Telecommunications Industry Association (TIA)
Type	National Standards Organization—Industry (Telecommunications)
Summary	TIA is a trade association representing the global information and communications technology industries through standards development and other activities for companies involved in telecommunications, broadband, mobile wireless, information technology, networks, cable, satellite, unified communications, emergency communications, and the greening of technology. Within the association, each area is represented by engineering committees and subcommittees that formulate standards to serve the industry and users.[164]
Relevant Engineering Committees	• TR-8 Mobile and Personal Private Radio Standards: Engineering Committee TR-8 formulates and maintains standards for private radio communications systems and equipment for both voice and data applications. TR-8 addresses all technical matters for systems and services, including definitions, interoperability, compatibility, and compliance requirements. The types of systems addressed by these standards include business and industrial dispatch applications, as well as public safety (such as police, ambulance and firefighting) applications.[165] • TR-42 Engineering Committee: TR-42 develops and maintains voluntary telecommunications standards for telecommunications cabling infrastructure in user-owned buildings, such as commercial buildings, residential buildings, homes, data centers, industrial buildings, etc. The generic cabling topologies, design, distances and outlet configurations as well as specifics for these locations are addressed. The committee's standards work covers requirements for copper and optical fiber cabling components (such as cables, connectors and cable assemblies), installation, and field testing in addition to the administration, pathways and spaces to support the cabling.[166]

---

[164]TIA, *About TIA*. Available at: http://www.tiaonline.org/about/ (last accessed February 26, 2016).

[165] TIA, *TR-8 Mobile and Personal Private Radio Standards*. Available at: http://www.tiaonline.org/all-standards/committees/tr-8 (last accessed February 26, 2016).

[166] TIA, *TR-42 Telecommunications Systems Standards*. Available at: http://www.tiaonline.org/all-standards/committees/tr-42 (last accessed February 26, 2016).

Relevant Engineering Committees (continued)	• TR-45 Mobile and Point-to-Point Communications Standards: Engineering Committee TR-45 develops performance, compatibility, interoperability, and service standards for mobile and personal communications systems. These standards pertain to, but are not restricted to, service information, wireless terminal equipment, wireless base station equipment, wireless switching office equipment, ancillary apparatus, auxiliary applications, inter-network and intersystem operations, interfaces, and wireless packet data technologies.[167]
	• TR-48 Vehicular Telematics: Engineering Committee TR-48 is responsible for development and maintenance of standards relating to vehicular telematics equipment and services. TR-48 works with other TIA committees, national and international standards organizations, and other relevant entities to ensure work items are necessary and not duplicative.[168]
	• TR-50 Smart Device Communications Standards: This committee is responsible for the development and maintenance of access agnostic interface standards for the monitoring and bi-directional communication of events and information between machine-to-machine (M2M) systems and smart devices, applications or networks. These standards development efforts pertain to, but are not limited to, the functional areas as noted: Reference Architecture, Informational Models and Standard Objects, Protocol Aspects, Software Aspects, Conformance and Testing, and Security.[169]
Standards	• TIA-102 Series: *Telecommunications, Land Mobile Communications,* (APCO/Project 25)
	• TIA TSB-102.BACC: *Project 25 Interface-RF-Subsystem Interface Overview*
	• TIA TSB-102.BAGA: *Project 25 Console Subsystem Interface Overview*
	• TIA TSB-102.BAJA: *Project 25 Location Services Overview*
	• TIA-102.BAED: *Project 25 Packet Data Logical Link Control Procedures*
	• TIA-TSB-146: *Telecommunications IP Telephony Infrastructures IP Telephony Support for Emergency Calling Service*
	• TIA-222 Revision G: *Structural Standard for Antenna Supporting Structures and Antennas*
	• TIA-568 Set: *Commercial Building Telecommunications Cabling Standard Set*
	• TIA-606 Revision B: *Administration Standard for Telecommunications Infrastructure*
	• TIA-664.529: *Wireless Features Description: Emergency Services (9-1-1)*

---

[167] TIA, *TR-45 Mobile and Point-to-Point Communications Standards*. Available at: http://www.tiaonline.org/all-standards/committees/tr-45 (last accessed February 26, 2016).

[168] TIA, *TR-48 Vehicular Telematics*. Available at: http://www.tiaonline.org/all-standards/committees/tr-48 (last accessed February 26, 2016).

[169]TIA, *TR-50 Smart Device Communications*. Available at: https://global.ihs.com/landing_page_tia.cfm?&rid=TIA&seg_code=TR-50&org_code=TIA (last accessed February 26, 2016).

Standards (continued)	• TIA-942 Revision A: *Telecommunications Infrastructure Standard for Data Centers* • TIA-1039: *QoS Signaling for IP QoS Support and Sender Authentication* • TIA-1057: *Telecommunications IP Telephony Infrastructure Link Layer Discovery Protocol for Media Endpoint Devices* • TIA-1191: *Callback to an Emergency Call Origination Stage 1 Requirements* • TIA-4973.211: *Requirements for the Mission Critical Priority and QoS Control Service* • TIA/EIA/IS-834: *G3G CDMA-DS to ANSI/TIA/EIA-41* • TIA J-STD-110: *Joint ATIS/TIA Native SMS to 9-1-1 Requirements and Architecture Specification Release 2* • TIA J-STD-110.01: *Joint ATIS/TIA Implementation Guide for J-STD-110, Joint ATIS/TIA Native SMS to 9-1-1 Requirements and Architecture Specification Release 2* • TIA J-STD-110.A: *Joint ATIS/TIA Supplement A to J-STD-110, Joint ATIS/TIA Native SMS to 9-1-1 Requirements & Architecture Specification*
Strategic Initiatives	The initiatives listed below are of high interest to the telecommunications community and are areas in which TIA has developed standards or closely monitors for future standards development needs: • Project 25 • CALEA • 3GPP2 • ITU [170]
Coordinated Activities	• 3GPP • 3GPP2 • APCO International • ATIS • ETSI • IETF • ITU-T • ANSI: TIA is an ANSI-accredited SDO[171]
Effects on NG911	• Develops standards adhered to by originating service providers' (OSP) network and applications services for emergency calling.
Website	http://www.tiaonline.org/

---

[170] TIA, *Strategic Initiatives*. Available at: http://www.tiaonline.org/standards/strategic-initiatives (last accessed February 26, 2016).

[171] TIA, *TIA Homepage*. Available at: http://www.tiaonline.org/ (last accessed February 26, 2016).

Wi-Fi Alliance

Name	Wi-Fi Alliance
Type	Industry Organization
Summary	The Wi-Fi Alliance is a global non-profit organization with the goal of driving adoption of a single worldwide standard for high-speed wireless local area networking.
Mission	The Wi-Fi Alliance's mission is to: • Foster highly effective collaboration among stakeholders • Deliver excellent connectivity experiences through interoperability • Embrace technology innovation • Promote the adoption of our technologies worldwide • Advocate for fair worldwide spectrum rules • Lead, develop and embrace industry-agreed standards[172]
Related Activities	• International Telecommunications Union (ITU)
Website	http://www.wi-fi.org/

---

[172] Wi-Fi Alliance, *Organization.* Available at: http://www.wi-fi.org/who-we-are (last accessed February 26, 2016).

WiMAX Forum

Name	WiMAX Forum
Type	Industry Organization
Summary	The WiMAX Forum is an industry-led, not-for-profit organization formed to certify and promote the compatibility and interoperability of broadband wireless products based upon the harmonized IEEE 802.16/ETSI HiperMAN standard.[173]
Mission	The WiMAX Forum is the worldwide consortium focused on global adoption of WiMAX and chartered to establish certification processes that achieve interoperability, publish technical specifications based on recognized standards, promote the technology, and pursue a favorable regulatory environment.
Coordinated Activities	• International Telecommunications Union (ITU)[174]
Website	http://www.wimaxforum.org/

---

[173] WiMAX Forum, *About the WiMAX Forum.* Available at: http://www.wimaxforum.org/about (last accessed February 26, 2016).
[174] WiMAX Forum, *WiMax in ITU.* Available at: http://www.wimaxforum.org/Default.aspx?PageID=13610805&A=SearchResult&SearchID=23048647&ObjectID=13610805&ObjectType=1 (last accessed February 26, 2016).

Moving Forward

It is important for NG911 stakeholders to be mindful of how the un-standardized, semi-planned approach to standards development can and will affect the ability of PSAPs and emergency response entities to effectively share information and be interoperable. To alleviate this issue, increased national activities (e.g., state oversight, state/regional compliant designs, and federal coordination working groups) should be considered to ensure that a complete set of NG911 open standards are accepted and adopted by all relevant stakeholders. This should include active participation by the stakeholders. Additionally, increased national collaboration could be utilized to monitor progress on the options below to address standards, technological barriers, and issues identified in A *National Plan for Migrating to IP-Enabled 9-1-1 Systems*:

- Strive for IP-enabled 911 open standards and understand future technology trends to encourage system interoperability and emergency data sharing
- Establish routing, prioritization, and business rules
- Determine the responsible entity and mechanisms for location acquisition and determination
- Establish system access and security controls to protect and manage access to the IP-enabled 911 system of systems
- Develop a certification and authentication process to ensure service providers and 911 authorities meet security and system access requirements[175]

Lastly, without processes and protocols (e.g., certification and authentication, routing business rules), the benefits of the NG911 system—including routing based on criteria beyond location and connection of service providers beyond common carriers to the 911 system—are unlikely to be fully realized.

A significant number and variety of standards potentially will have a key impact on the implementation of NG911. Continuing to actively monitor standards that have been completed, along with relevant standards that are likely to emerge, will be essential in ensuring the greatest benefit to the global community. The National 911 Program will continue to monitor NG911 standards and update this "living" document to reflect the progress made by SDOs and SSOs.

---

[175]National 911 Program, A *National Plan for Migrating to IP-Enabled 9-1-1 Systems.* Available at: http://911.gov/911-issues/standards.html (last accessed February 25, 2016).

Acronym List

ACRONYM	DESCRIPTION
3GPP	3rd Generation Partnership Project
AACN	Advanced Automatic Collision Notification
AES	Advanced Encryption Standard
AIN	Advanced Intelligent Network
ALI	Automatic Location Identification
AMF	Access Measurement Function
ANS	American National Standard
ANSI	American National Standards Institute
APCO	Association of Public-Safety Communication Officials, International
API	Application Programming Interface
AQS	ALI Query Service
ARIB	Association of Radio Industries and Businesses
ASAP	Automated Secure Alarm Protocol
ASD	ANSI-accredited Standards Developer
ATIS	Alliance for Telecommunications Industry Solutions
BBF	Broadband Forum
BCF	Border Control Function
BES	Bulk Electric System
BFD	Bidirectional Forwarding Detection
BGP	Border Gateway Protocol
BICSI	Building Industries Consulting Service International
BIM	Building Information Modeling
BJA	Bureau of Justice Assistance
BNG	Broadband Network Gateway
BSS	Base Station System
BSS – MSC	Base Station System – Mobile-services Switching Center
BWA	Broadband Wireless Access
CAD	Computer Aided Dispatch
CAI	Common Air Interface
CALEA®	Commission on Accreditation for Law Enforcement Agencies, Inc.
CAP	Common Alerting Protocol
CCSA	China Communications Standards Association
CDMA	Code Division Multiple Access
CEMA	Connection Establishment for Media Anchoring
CISA	Certified Information Systems Auditor

ACRONYM	DESCRIPTION
CityGML	City Geography Markup Language
CJI	Criminal Justice Information
CJIS	Criminal Justice Information Services
CLDXF	Civic Location Data Exchange Format
CMAS	Commercial Mobile Alert Service
CMRS	Commercial Mobile Radio Service
CMSP	Commercial Mobile Service Provider
CN	Core Network
COGO	Coalition of Geospatial Organizations
COMEDIA	Connection-oriented Media
COS	Class of Service
CPE	Customer Premise Equipment
CPP	Common Profile Presence
CS&C	Office of Cybersecurity and Communications
CSRIC	Communications Security, Reliability, and Interoperability Council
CSSI	Console Subsystem Interface
CTO	Communications Training Officer
DAS	Distributed Antenna System
DHCP	Dynamic Host Control Protocol
DHHS	Department of Health and Human Services
DHS	Department of Homeland Security
DNS	Domain Name System
DOC	Department of Commerce
DOJ	Department of Justice
DOT	Department of Transportation
DS	Differentiated Services
DSCP	Differentiated Code Point
DSL	Digital Subscriber Line
DSS	Data Security Standard
E911	Enhanced 911
EAAC	Emergency Access Advisory Committee
ECES	Entities Consuming Emergency Services
eCNAM	Enhanced Calling Name
ECRF	Emergency Call Routing Function
ECRIT	Emergency Context Resolution with Internet Technologies
ECS	Emergency Calling Service
EDGE	Enhanced Data Rates for GSM Evolution
EDXL	Emergency Data Exchange Language

ACRONYM	DESCRIPTION
EDXL-DE	EDXL-Distribution Element
EDXL-RM	EDXL-Resource Messaging
EDXL-TEC	EDXL-Tracking of Emergency Clients
EDXL-TEP	EDXL-Tracking of Emergency Patients
EFD	Emergency Fire Dispatch
eHRPD	Evolved High Rate Packet Data
EIA	Electronics Industry Alliance
EIDD	Emergency Incident Data Document
EISI	Emergency Information Services Interface
EMD	Emergency Medical Dispatch
EM-TC	Emergency Management Technical Committee
EMTEL	Emergency Communications
ENUM	E.164 Number Mapping
EP	Emergency Preparedness
EPC	Evolved Packet Core
EPD	Emergency Police Dispatch
EPES	Entities Providing Emergency Services
ERIC	Emergency Response Interoperability Center
ESC	Executive Steering Council
ESIF	Emergency Services Interconnection Forum
ESInet	Emergency Services IP Network
ESM	Emergency Services & Methodologies
ESMI	Emergency Services Messaging Interface
ESNet	Emergency Services Network
ES-NGN	Emergency Services Next Generation Network
ESNI	Emergency Services Network Interfaces
ESQK	Emergency Services Query Key
ESRD	Emergency Services Routing Digit
ESRK	Emergency Services Routing Key
ESS	Electronic Safety and Security
ESW	Emergency Services Workshop
ETC	Emergency Telecommunicator Certification
ETS	Emergency Telecommunications Service
ETSI	European Telecommunications Standards Institute
FCC	Federal Communications Commission
FDD	Frequency Division Duplex
FGDC	Federal Geographic Data Committee
FIPS	Federal Information Processing Standard

ACRONYM	DESCRIPTION
FIPS PUB	FIPS Publication
FLAP	Flexible LDF-AMP Protocol
GEOPRIV	Geographic Location/Privacy
GETS	Government Emergency Telecommunications Service
GIS	Geographic Information System
GML	Geography Markup Language
GPON	Gigabit Passive Optical Network
GPRS	General Packet Radio Service
GRA	Government and Regulatory Agency
GSM	Global System for Mobile Communications
HAVE	Hospital Availability Exchange
HDSSC	Homeland Defense and Security Standardizations Collaborative
HELD	HTTP-enabled Location Delivery
HMI	Human Machine Interface
HRPD	High Rate Packet Data
HSGW	eHRPD Serving Gateway
HSSP	Homeland Security Standards Panel
HTTP	Hypertext Transfer Protocol
I2F	Information Interoperability Framework
IACP	International Association of Chiefs of Police
IAED	International Academies of Emergency Dispatch
ICE	Industry Collaboration Event
ICO	Implementation and Coordination Office
ICT	Information and Communications Technology
IEC	International Electrotechnical Commission
IEEE	Institute of Electrical and Electronics Engineers
IETF	Internet Engineering Task Force
IJIS	Integrated Justice Information Systems
IM	IP Multimedia
IMIS	Incident Management Information Sharing
IMS	IP Multimedia Subsystem
IMSI	International Mobile Subscriber Identity
INP	Interim Number Portability
IOT	Internet of Things
IP	Internet Protocol
IPAWS	Integrated Public Alert and Warning System
IPR	Intellectual Property Rights
IPTV	IP Television

ACRONYM	DESCRIPTION
ISDN	Integrated Services Digital Network
ISE	Information Sharing Environment
ISF	Information Security Forum
ISMS	Information Security Management Systems
ISO	International Organization for Standardization
ISSI	Inter-RF Subsystem Interface
ISUP	ISDN User Part
IT	Information Technology
ITL	Information Technology Laboratory
ITS	Intelligent Transportation Systems
ITS	Institute for Telecommunication Sciences
ITU	International Telecommunication Union
ITU-R	ITU—Radiocommunication Sector
ITU-T	ITU—Standardization Sector
IWS	Intelligent Workstation System
JPO	Joint Program Office
kHz	Kilohertz
LAN	Local Area Network
LCP	Location Configuration Protocol
LDF	Location Determination Function
LIS	Location Information Server
LLC	Logical Link Control
LLDP	Link Layer Discovery Protocol
LMR	Land Mobile Radio
LNP	Local Number Portability
LoST	Location-to-Service Translation
LTE	Long-term Evolution
LVF	Location Validation Function
M2M	Machine-to-machine
MAC	Media Access Control
MAN	Metropolitan Area Network
MAP	Mobile Application Part
MDA®	Model Driven Architecture®
MED	Media Endpoint Devices
MGCI	Media Gateway Control Interface
MGCP	Media Gateway Control Protocol
MHz	Megahertz
MIB	Management Information Base

ACRONYM	DESCRIPTION
MLP	Mobile Location Protocol
MLTS	Multi-line Telephone System
MMS	Multimedia Messaging Service
MOS	Mean Opinion Score
MOU	Memorandum of Understanding
MPC	Mobile Positioning Center
MS	Mobile Station
MS – BSS	Mobile Station – Base Station System
MSAG	Master Street Address Guide
MSC	Mobile-services Switching Center
MSRP	Message Session Relay Protocol
NBAC	NIEM Business Architecture Committee
NCMEC	National Center for Missing and Exploited Children
NE	Network Element
NEC	National Electrical Code®
NENA	National Emergency Number Association
NERC	North American Electric Reliability Corporation
NFPA	National Fire Protection Association
NG911	Next Generation 911
NGES	Next Generation Emergency Services
NGIIF	Next Generation Interconnection Interoperability Forum
NGN	Next Generation Network
NGP	Next Generation Protocols
NGPP	Next Generation Partner Program
NHTSA	National Highway Traffic Safety Administration
NIEM	National Information Exchange Model
NIST	National Institute of Standards and Technology
NNI	Network to Network Interface
NOBLE	National Organization of Black Law Enforcement Executives
NPPD	National Protection and Programs Directorate
NPSBN	Nationwide Public Safety Broadband Network
NRIC	Network Reliability and Interoperability Council
NS	National Security
NSA	National Sheriffs' Association
NSDI	National Spatial Data Infrastructure
NTAC	NIEM Technical Architecture Committee
NTIA	National Telecommunications and Information Administration
NTSIC	National Strategy for Trusted Identities in Cyberspace

ACRONYM	DESCRIPTION
OASIS	Organization for the Advancement of Structured Information Standards
OEC	Office of Emergency Communications
OGC®	Open Geospatial Consortium
OIC	Office of Interoperability and Compatibility
OJP	Office of Justice Programs
OMA	Open Mobile Alliance
OMB	Office of Management and Budget
OMG®	Object Management Group®
OpenLS	OpenGIS Location Service
OSP	Originating Service Provider
OSPF	Open Shortest Path First
OSS	Operations Support System
OST-R	Office of the Assistant Secretary for Research and Technology
pANI	Pseudo Automatic Number Identification
PCI	Payment Card Industry
PDE	Position Determining Equipment
PERF	Police Executive Research Forum
PIDF	Presence Information Data Format
PIDF-LO	Presence Information Data Format-Location Object
PML	Physical Measurement Laboratory
PMO	Program Management Office
PRACK	Provisional Response Acknowledgement
PSAP	Public Safety Answering Point
PSHSB	Public Safety and Homeland Security Bureau
PSTN	Public Switched Telephone Network
PTSC	Packet Technologies and Systems Committee
PTT	Push-to-talk
QA	Quality Assurance
QAE	Quality Assurance Evaluator
QI	Quality Improvement
QoS	Quality of Service
RF	Radio Frequency
RFAI	Request for Assistance Interface
RFC	Request for Comment
RFI	Request for Information
RG	Residential Gateway
RITA	Research and Innovative Technology Administration
RNA	Routing Number Authority

ACRONYM	DESCRIPTION
RTP	Real-time Transport Protocol
S&T	Science & Technology Directorate
SAFECOM	Wireless Public Safety Interoperable Communications Program
SBC	Session Border Controller
SCC	Standards Coordinating Council
SCTE	Society of Cable Telecommunications Engineers
SDO	Standards Development Organization
SDP	Session Description Protocol
SEC	Security
SHS	Secure Hash Standard
SIP	Session Initiated Protocol
SIPREC	SIP Recording
SMS	Short Message Service
SNMP	Simple Network Management Protocol
SOP	Standard Operating Procedure
SPO	Special Programs Office
SR	Selective Router
SR	Selective Router
SRIC	Standards Review and Interpretation Committee
SS7	Signaling System 7
SSO	Standards Setting Organization
SUPL	Secure User Plan Location
TCC	Text Control Center
TDD	Time Division Duplex
TDM	Time Division Multiplexing
TERT	Telecommunicator Emergency Response Taskforce
TFOPA	Task Force on Optimal PSAP Architecture
TIA	Telecommunications Industry Alliance
TISPAN	Telecommunications & Internet Converged Services & Protocols for Advanced Networks
TLS	Transport Layer Security
TMOC	Telecom Management and Operations Committee
TSAG	Transportation Safety Advancement Group
TSB	Technical Service Bulletin
TSDSI	Telecommunications Standards Development Society, India
TSG	Technical Specification Group
TTA	Telecommunications Technology Association, Korea
TTC	Telecommunication Technology Committee, Japan
TTY/TDD	Teletypewriter/Telecommunications Device for the Deaf

ACRONYM	DESCRIPTION
TVRA	Threat Vulnerability Risk Analysis
U.S.	United States
UA	User Agents
UMA	Universal Mobile Access
UML®	Unified Modeling Language®
UMTS	Universal Mobile Telecommunications System
URI	Uniform Resource Identifier
URISA	Urban and Regional Information Systems Association
URL	Uniform Resource Locator
URN	Uniform Resource Number
USM	User-based Security Model
UTRA	UTMS Terrestrial Radio Access
VACM	View-based Access Control Model
VoDSL	Voice over Digital Subscriber Line
VoIP	Voice over Internet Protocol
VOP	Voice over Packet
VPN	Virtual Private Network
WLAN	Wireless Local Area Network
WMS	Web Map Service
WSP	Wireless Service Provider
WTSC	Wireless Technologies and Systems Committee
XML	eXtensible Markup Language

Appendix A: Standards and Best Practices

Entity	Standard or Document ID	Standard or Document Title	Standard or Document Description	Associated Documents	Latest Revision/ Release Date	Standard or Document Type	Relation to NENA i3 Architecture				
							Client (C)	Access Networks (A)	Origination Networks (O)	ESInets (E)	PSAPs (P)
3GPP	3GPP TS 23.167 (Free)	IP Multimedia Subsystem (IMS) emergency sessions	Defines the service description (Stage 2) for emergency services in the IMS, including the elements necessary to support SIP multimedia emergency services.	ETSI TS 123 167	Version 13.1.0 December 15, 2015	Technical Standard (Product/ Design)		A	O		
3GPP	3GPP TS 23.228 (Free)	IP Multimedia Subsystem (IMS); Stage 2	Defines the Stage 2 service description for the IMS, which includes the elements necessary to support IP multimedia (IM) services.		Version 13.4.0 September 22, 2015	Technical Standard		A	O		
3GPP	3GPP TS 23.517 (Free)	TISPAN; IP Multimedia Subsystem (IMS); Functional architecture	Describes the IMS core component of the TISPAN NGN functional architecture and its relationships to other subsystems and components.	ETSI ES 282 007	Version 8.0.0 December 11, 2007	Technical Standard (Interface/ Design)		A	O		
3GPP	3GPP TS 24.229 (Free)	IP multimedia call control protocol based on Session Initiation Protocol (SIP) and Session Description Protocol (SDP); Stage 3	Defines a call control protocol for use in the IM Core Network (CN) subsystem based on the SIP and the associated SDP.		Version 13.4.0 December 18, 2015	Technical Standard		A	O		

Next Generation 911 (NG911) Standards Identification and Review

Entity	Standard or Document ID	Standard or Document Title	Standard or Document Description	Associated Documents	Latest Revision/ Release Date	Standard or Document Type	Relation to NENA i3 Architecture				
							Client (C)	Access Networks (A)	Origination Networks (O)	ESInets (E)	PSAPs (P)
3GPP	3GPP TS 29.010 (Free)	Information element mapping between Mobile Station - Base Station System (MS - BSS) and Base Station System - Mobile-services Switching Centre (BSS - MSC); Signaling Procedures and the Mobile Application Part (MAP)	Provides a detailed specification for the interworking between information elements contained in layer 3 messages sent on the MS-MSC interface where the MSC acts as a transparent relay of information; provides a detailed specification for the interworking between information elements contained in BSSMAP messages sent on the BSC-MSC interface and parameters contained in MAP services sent over the MSC-VLR interface where the MSC acts as a transparent relay of information.		Version 13.0.0 December 17, 2015	Technical Standard		A	O		
3GPP	3GPP TSG SA Release 12 (Free)	Release 12	Focuses on the use of LTE technology for emergency and security services, with technical specifications for mission-critical application layer functional elements and interfaces being developed in the newly formed SA6 working group.		March 2015	Technical Standard		A			

Next Generation 911 (NG911) Standards Identification and Review

Entity	Standard or Document ID	Standard or Document Title	Standard or Document Description	Associated Documents	Latest Revision/ Release Date	Standard or Document Type	Relation to NENA i3 Architecture				
							Client (C)	Access Networks (A)	Origination Networks (O)	ESInets (E)	PSAPs (P)
3GPP	3GPP TSG SA Release 13 (Free)	Release 13	Exploits new business opportunities such as public safety and critical communications, explores Wi-Fi integration and system capacity and stability.		In progress; planned release March 2016	Technical Standard		A			
3GPP	3GPP TSG SA Release 14 (Free)	Release 14	Supports V2x services, eLAA, 4 band carrier aggregation, and inter-band carrier aggregation.		In progress; planned release June 2017	Technical Standard		A			
3GPP2	3GPP2 S.R0006-529-A (Free)	Wireless Features Description: Emergency Services	Describes the wireless emergency services (i.e., 9-1-1) feature that permits a subscriber to dial 9-1-1 and be connected to a PSAP (appropriate to the calling subscriber's current location) to request an emergency response from the appropriate agency (e.g., fire, police, ambulance).		Version 1.0 June 2007	Technical Standard (Product/ Design)		A	O		
3GPP2	3GPP2 X.S0049-0 (Free)	All-IP Network Emergency Call Support	Describes the service and procedures in the IMS, including the elements necessary to support emergency services in IMS.		Version 1.0 February 18, 2008	Technical Standard (Interface/ Design)		A	O		

Entity	Standard or Document ID	Standard or Document Title	Standard or Document Description	Associated Documents	Latest Revision/ Release Date	Standard or Document Type	Relation to NENA i3 Architecture					
							Client (C)	Access Networks (A)	Origination Networks (O)	ESInets (E)	PSAPs (P)	
3GPP2	3GPP2 X.S0057-A (Free)	E-UTRAN - eHRPD Connectivity and Interworking: Core Network Aspects	Provides a specification of the functions and interfaces of the eHRPD Serving Gateway (HSGW) and the IP-level interfaces of the eHRPD user equipment.		Version 2.0 October 2012	Technical Standard		A	O			
3GPP2	3GPP2 X.S0060-0 (Free)	HRPD Support for Emergency Services	Describes the characteristics for the provisioning of IMS emergency services using the HRPD network.		Version 1.0 July 2008	Technical Standard (Product/ Design)		A	O			
APCO	APCO/NENA ANS 1.107.1.2015 (Free)	Standard for the Establishment of a Quality Assurance and Quality Improvement Program for Public Safety Answering Points	Provides QA and improvement guidelines.		Version 1 April 2, 2015	Operational Standard					P	
APCO	APCO ANS 1.116.1-2015 (Free)	Public Safety Communications Common Status Codes for Data Exchange	Provides a standardized list of status codes that can be used by emergency communications and public safety stakeholders when sharing incident related information; each agency should map their internal codes to the standardized list.		Version 1 April 7, 2015	Operational Standard				E	P	

Entity	Standard or Document ID	Standard or Document Title	Standard or Document Description	Associated Documents	Latest Revision/ Release Date	Standard or Document Type	Relation to NENA i3 Architecture				
							Client (C)	Access Networks (A)	Origination Networks (O)	ESInets (E)	PSAPs (P)
APCO	APCO ANS 1.112.1-2014 (Free)	Best Practices for The Use of Social Media in Public Safety Communications	Provides a consistent foundation for agencies to develop specific operational procedures and competencies; recognizes the need for each agency to customize specific procedures to their local environment.		Version 1 2014	Operational Standard					P
APCO	APCO ANS 1.110.1-2015 (Free)	Multi-Functional Multi-Discipline Computer Aided Dispatch (CAD) Minimum Functional Requirements	Provides minimum functional requirements that a CAD system shall include, broken down by public safety discipline; also identified are the optional functional requirements that a CAD system should include.		Version 1 January 9, 2015	Operational Standard					P
APCO	APCO/NPSTC ANS 1.104.1-2010 (Free)	Standard Channel Nomenclature for the Public Safety Interoperability Channels	Provides standard nomenclature for FCC and NTIA-designated nationwide interoperability channels used for public safety voice communications.		Version 1 June 9, 2010 (Version 2 in Development)	Operational Standard					P

Entity	Standard or Document ID	Standard or Document Title	Standard or Document Description	Associated Documents	Latest Revision/Release Date	Standard or Document Type	Relation to NENA i3 Architecture				
							Client (C)	Access Networks (A)	Origination Networks (O)	ESInets (E)	PSAPs (P)
APCO	APCO ANS 1.101.3-2015 (Free)	Standard for Public Safety Telecommunicators When Responding to Calls of Missing, Abducted and Sexually Exploited Children	Presents the missing, abducted, and/or sexually exploited child response process for public safety telecommunicators; includes the process from first response through ongoing incident and case support.		Version 3 January 8, 2015	Operational Standard					P
APCO	APCO/NENA ANS 1.105.2-2015 (Free)	Standard for Telecommunicator Emergency Response Taskforce (TERT) Deployment	Includes information to provide guidance and helpful material regarding the development, maintenance, and deployment of a TERT.		Version 2 July 14, 2015	Operational Standard					P
APCO	APCO ANS 3.103.2-2013 (Free)	Wireless 9-1-1 Deployment and Management Effective Practices Guide	Provides effective practices to increase a PSAP manager's understanding of the technology application and the ability to better manage wireless calls, as well as public and responder expectations.		Version 2 September 27, 2013	Operational Standard					P
APCO	APCO ANS 1.111.1-2013 (Free)	Public Safety Communications Common Disposition Codes for Data Exchange	Provides a standardized list of disposition codes to facilitate effective incident exchange between NG9-1-1 PSAPs and other authorized agencies.		Version 1 December 12, 2013	Operational Standard					P

Next Generation 911 (NG911) Standards Identification and Review

Entity	Standard or Document ID	Standard or Document Title	Standard or Document Description	Associated Documents	Latest Revision/ Release Date	Standard or Document Type	Relation to NENA i3 Architecture				
							Client (C)	Access Networks (A)	Origination Networks (O)	ESInets (E)	PSAPs (P)
APCO	APCO/CSAA ANS 2.101.2-2014 (Free)	Alarm Monitoring Company to Public Safety Answering Point (PSAP) Computer-Aided Dispatch (CAD) Automated Secure Alarm Protocol (ASAP)	Provides detailed technical data to software providers who support CAD systems or alarm monitoring applications concerning the common data elements and structure that shall be utilized when electronically transmitting a new alarm event from an alarm monitoring company to a PSAP.		Version 2 August 5, 2014	Technical Standard					P
APCO	APCO ANS 2.103.1-2012 (Free)	Public Safety Communications Common Incident Types For Data Exchange	Defines and outlines public safety communications common incident types for data exchange.		Version 1 November 2012	Technical Standard				E	P
APCO	APCO ANS 3.101.2-2013 (Free)	Core Competencies and Minimum Training Standards for Public Safety Communications Training Officer (CTO)	Addresses the minimum training requirements necessary to foster levels of consistency for all personnel in an emergency communications environment assigned to providing on-the-job training to active 9-1-1 professionals and telecommunicators, as well as to promote the leadership role of the CTO.		Version 2 January 2013	Training Standard					P

Entity	Standard or Document ID	Standard or Document Title	Standard or Document Description	Associated Documents	Latest Revision/ Release Date	Standard or Document Type	Relation to NENA i3 Architecture				
							Client (C)	Access Networks (A)	Origination Networks (O)	ESInets (E)	PSAPs (P)
APCO	APCO ANS 3.108.1.2014 (Free)	*Core Competencies and Minimum Training Standards for Public Safety Communications Instructor*	Defines the minimum training standards for PSAP instructors.		Version 1 February 3, 2014	Training Standard					P
APCO	APCO ANS 3.106.1-2013 (Free)	*Core Competencies and Minimum Training Standards for Public Safety Communications Quality Assurance Evaluators (QAE)*	Defines the minimum training standards for PSAP QA evaluators.		Version 1 April 11, 2013	Training Standard					P
APCO	APCO ANS 3.102.1-2012 (Free)	*Core Competencies and Minimum Training Standards for Public Safety Communications Supervisor*	Identifies the core competencies and minimum training requirements for public safety communications supervisors relating to managing daily operations, performing administrative duties, and maintaining employee relations.		Version 1 December 7, 2012	Training Standard					P
APCO	APCO ANS 3.109.2.2014 (Free)	*Core Competencies and Minimum Training Standards for Public Safety Communications Manager/Director*	Defines the core competencies and minimum training requirements for communications managers and/or directors.		Version 2 June 9, 2014	Training Standard					P

Next Generation 911 (NG911) Standards Identification and Review

Entity	Standard or Document ID	Standard or Document Title	Standard or Document Description	Associated Documents	Latest Revision/ Release Date	Standard or Document Type	Relation to NENA i3 Architecture				
							Client (C)	Access Networks (A)	Origination Networks (O)	ESInets (E)	PSAPs (P)
APCO	APCO ANS 3.104.1-2012 (Free)	Core Competencies and Minimum Training Standards for Public Safety Communications Training Coordinator	Defines the minimum training standards for PSAP training coordinators.		Version 1 December 7, 2012	Training Standard					P
APCO	APCO ANS 3.103.2.2015 (Free)	Minimum Training Standards for Public Safety Telecommunicators	Identifies the minimum training requirements for public safety telecommunicators, which typically includes with receiving, processing, transmitting, and conveying public safety information to dispatchers, first responders (police, fire, EMS), and emergency management personnel.		Version 2 July 14, 2015	Training Standard					P
APCO	APCO ANS 3.107.1.2015 (Free)	Core Competencies and Minimum Training Requirements for Public Safety Communications Technician	Defines the minimum training standards for PSAP communications technicians.		Version 1 February 24, 2015	Training Standard					P

Next Generation 911 (NG911) Standards Identification and Review

Entity	Standard or Document ID	Standard or Document Title	Standard or Document Description	Associated Documents	Latest Revision/ Release Date	Standard or Document Type	Relation to NENA i3 Architecture				
							Client (C)	Access Networks (A)	Origination Networks (O)	ESInets (E)	PSAPs (P)
APCO	APCO/NENA ANS 3.105.1-2015 (Free)	Minimum Training Standard for TTY/TDD Use in the Public Safety Communications Center	Defines the minimum training standards for TTY/TDD use in communications centers.		Version 1 February 24, 2015	Training Standard					P
APCO	APCO/NENA ANS 1.102.2-2010 (Free)	Public Safety Answering Point (PSAP) Service Capability Criteria Rating Scale	Provides an assessment tool for PSAP managers and their governing authorities to identify their current level of service capability; objectively assesses the capabilities of the PSAP against models representing the best level of preparedness, survivability, and sustainability amidst a wide range of natural and man-made events.		Version 2 July 28, 2010 (Version 3 in Development)	Operational Standard					P
APCO	APCO 1.108.1-201x	Minimum Operational Standards for the Use of TTY/TDD devices in the Public Safety Communications Center	Defines the minimum operational standards for the use of TTY/TDD devices in a PSAP.		In Development	Operational Standard					P
APCO	APCO 1.113.1-201x	Public Safety Communications Call Handling Process	Provides best practices for call handling in the PSAP.		In development	Operational Standard					P

Next Generation 911 (NG911) Standards Identification and Review

Entity	Standard or Document ID	Standard or Document Title	Standard or Document Description	Associated Documents	Latest Revision/ Release Date	Standard or Document Type	Relation to NENA i3 Architecture				
							Client (C)	Access Networks (A)	Origination Networks (O)	ESInets (E)	PSAPs (P)
APCO	APCO 1.114.1-201x	Vehicle Telematics Best Practices	Provides best practices for vehicle telematics in the PSAP.		In Development	Operational Standard					P
APCO	APCO 1.115.1-201x	Core Competencies, Operational Factors, and Training for Next Generation Technologies in Public Safety Communications	Identifies the core competencies, operational factors and minimum training requirements relating to next generation technologies.		In Development	Operational Standard					P
APCO	APCO 2.102.1-201x	Advanced Automatic Collision Notification (AACN) Data Set	Describes and outlines the AACN data set.		In Development	Technical Standard				E	P
APCO	APCO 2.104.1-201x	Application Integration (for/with) Public Safety Answering Points (PSAPs) and Public Safety Responders			In Development	Technical Standard				E	P
APCO	APCO/NENA 2.105.1-201x	NG9-1-1 Emergency Incident Data Document (EIDD)	Provides format for sharing emergency incident information.		In Development	Technical Standard				E	P
ATIS	ATIS-0100022 (Fee/Charge)	Priority Classification Levels for Next Generation Networks	Formalizes a set of priority classification levels for admission control and service restoration in NGNs; highest priority classifications are reserved for ETS.		December 2008	Technical Standard		A	O	E	

Next Generation 911 (NG911) Standards Identification and Review

Entity	Standard or Document ID	Standard or Document Title	Standard or Document Description	Associated Documents	Latest Revision/ Release Date	Standard or Document Type	Relation to NENA i3 Architecture				
							Client (C)	Access Networks (A)	Origination Networks (O)	ESInets (E)	PSAPs (P)
ATIS	ATIS-0300104 (Fee/Charge)	Next Generation Interconnection Interoperability Forum (NGIIF) NGN Reference Document - NGN Basics, Emergency Services, NGN Testing, and Network Survivability	Provides basic information regarding NGNs, as applicable to the NGIIF.	ATIS-0300109, ATIS-0300112, ATIS-0300111	September 2015	Technical Standard		A	O	E	
ATIS	ATIS-0500001 (Fee/Charge)	High Level Requirements for Accuracy Testing Methodologies	Provides a common frame of reference that stakeholders can use to validate the accuracy methodology of 9-1-1 location technologies and whether test equipment meets requirements.		November 2011	Technical Report		A	O		
ATIS	ATIS-0500002.2008(R2013) (Fee/Charge)	Emergency Services Messaging Interface (ESMI)	Contains standards for an Emergency Services Interface to the Emergency Services Network (ESNet); specifies protocols and message sets for use in the ESMI.		July 2008	Technical Standard (Interface/ Design)		A	O		

Entity	Standard or Document ID	Standard or Document Title	Standard or Document Description	Associated Documents	Latest Revision/ Release Date	Standard or Document Type	Relation to NENA i3 Architecture				
							Client (C)	Access Networks (A)	Origination Networks (O)	ESInets (E)	PSAPs (P)
ATIS	ATIS-0500003 (Fee/Charge)	Routing Number Authority (RNA) for pseudo Automatic Number Identification Codes (pANIs) Used for Routing Emergency Calls: pANI Assignment Guidelines and Procedures	Contains the guidelines and procedures for the assignment and use of pANIs used to route emergency calls, such as E9-1-1 calls or other types of emergency calls that need to become native E9-1-1 calls throughout the North American E9-1-1 systems (U.S. and Canada).		July 2005	Technical Standard		A	O		
ATIS	ATIS-0500004 (Fee/Charge)	Recommendation for the Use of Confidence and Uncertainty for Wireless Phase II	Contains ESIF recommendation for managing location confidence and uncertainty for wireless Phase 2 calls.		August 2005	Technical Standard		A	O		P
ATIS	ATIS-0500005 (Fee/Charge)	Standard Wireless Text Message Case Matrix	Addresses the need for standard wireless text messages; some PSAP screen formats provide space ALI text messages and the text messages are used to alert the call taker of a unique condition.		September 2005	Technical Standard		A	O		P

Next Generation 911 (NG911) Standards Identification and Review

Entity	Standard or Document ID	Standard or Document Title	Standard or Document Description	Associated Documents	Latest Revision/ Release Date	Standard or Document Type	Relation to NENA i3 Architecture Client (C)	Access Networks (A)	Origination Networks (O)	ESInets (E)	PSAPs (P)
ATIS	ATIS-0500006.2008(R2013) (Fee/Charge)	Emergency Information Services Interfaces (EISI) ALI Service	Specifies protocols and message sets used within the ESNet to communicate between Entities Consuming Emergency Services (ECES) and Entities Providing Emergency Services (EPES).		August 2008	Technical Standard (Interface-Data/Design)		A	O		
ATIS	ATIS-0500007.2008 (Fee/Charge)	Emergency Information Services Interface (EISI) Implemented with Web Services	Specifies protocols and message sets used within the ESNet to communicate via web services between ECES and EPES.		January 2008	Technical Standard (Interface-Data/Design)		A	O		
ATIS	ATIS-0500008 (Fee/Charge)	Emergency Services Network Interfaces (ESNI) Framework	Defines the framework and structure of the ESNI suite of standards; includes the ESMI that provides interconnections between next generation PSAPs and the ESNet.		October 2006	Technical Report		A	O		
ATIS	ATIS-0500009 (Fee/Charge)	High Level Requirements for End-to-End Functional Testing	Establishes procedures/standards to test that delivery of wireless 9-1-1 data remains constant through the network and is delivered with integrity to the PSAP.		April 2006	Technical Report		A	O	E	P

Entity	Standard or Document ID	Standard or Document Title	Standard or Document Description	Associated Documents	Latest Revision/ Release Date	Standard or Document Type	Relation to NENA i3 Architecture				
							Client (C)	Access Networks (A)	Origination Networks (O)	ESInets (E)	PSAPs (P)
ATIS	ATIS-0500013 (Fee/Charge)	Approaches to Wireless E9-1-1 Indoor Location Performance Testing	Provides recommendations for indoor wireless testing methodologies and validation.		February 2010	Technical Standard		A	O		
ATIS	ATIS-0500015.2010 (Fee/Charge)	Flexible LDF-AMF (Location Determination Function – Access Measurement Function) Protocol (FLAP) Specification	Provides a framework and associated protocols to allow an LDF to obtain the value of relevant network parameters associated with an end device, and from which the location of that end device may be determined.		August 2010	Technical Standard		A	O		
ATIS	ATIS-0500017 (Fee/Charge)	Considerations for an Emergency Services Next Generation Network (ES-NGN)	Defines an emergency services architecture based upon the ATIS definition of an ES-NGN; identifies potential standards gaps and focuses on the interconnection between the ES-NGN and networks that originate emergency calls.		June 2009	Technical Report		A	O		

Next Generation 911 (NG911) Standards Identification and Review

Entity	Standard or Document ID	Standard or Document Title	Standard or Document Description	Associated Documents	Latest Revision/ Release Date	Standard or Document Type	Relation to NENA i3 Architecture				
							Client (C)	Access Networks (A)	Origination Networks (O)	ESInets (E)	PSAPs (P)
ATIS	ATIS-0500018 (Fee/Charge)	P-ANI Allocation Tables for ESQKs, ESRKs, and ESRDs	Contains ESQK, ESRK, and ESRD allocation tables and capacities; assists Wireless Service Providers (WSPs) and Mobile Positioning Centers (MPCs) in improving the efficacy of p-ANI number use and administration, and complement preservation and utilization of limited p-ANI number resources.		August 2014	Technical Standard		A	O		
ATIS	ATIS-0500019.2010 (Fee/Charge)	Request for Assistance Interface (RFAI) Specification	Defines/describes the RFAI between the ES-NGN and a PSAP.		September 2010	Technical Standard				E	P
ATIS	ATIS-0500021 (Fee/Charge)	Supplemental Location Data	Contains standard for including supplemental location data to the ALI database from technologies providing indoor radio frequency (RF) coverage requiring a small signal footprint.		October 2012	Technical Standard		A	O	E	P
ATIS	ATIS-0500022 (Fee/Charge)	Test Plan Input for a Location Technology Test Bed	Leverages earlier standards and methods to provide a broad baseline test plan document for wireless indoor location accuracy testing.	ATIS-050000, 0500001, 0500013, CSRIC III WG3	October 2012	Technical Standard		A	O		P

Next Generation 911 (NG911) Standards Identification and Review

Entity	Standard or Document ID	Standard or Document Title	Standard or Document Description	Associated Documents	Latest Revision/ Release Date	Standard or Document Type	Relation to NENA i3 Architecture				
							Client (C)	Access Networks (A)	Origination Networks (O)	ESInets (E)	PSAPs (P)
ATIS	ATIS-0500023 (Fee/Charge)	Applying Common IMS to NG9-1-1 Networks	Provides the stage 1 definition for an IMS-based next generation emergency services architecture based on the 3GPP IMS standards.		April 2013	Technical Standard		A	O	E	
ATIS	ATIS-0500024 (Fee/Charge)	Comparison of SIP Profiles	Compares SIP profiles defined by ATIS, 3GPP, and NENA as they relate to emergency services.		April 2013	Technical Report		A	O		
ATIS	ATIS-0500025 (Fee/Charge)	Class of Service Support for Semi-Static Wireless	Addresses E9-1-1 Class of Service associated with a small cell that has a less than 100 meter coverage in an indoor environment.		July 2013	Technical Standard		A	O	E	P
ATIS	ATIS-0500026 (Fee/Charge)	Operational Impacts on Public Safety of ATIS-0700015, Implementation of 3GPP Common IMS Emergency Procedures for IMS Origination and ESInet/Legacy Selective Router Termination	Explains the IP to NG9-1-1 interfaces, without overdependence on technical terms and acronyms, to assist public safety in understanding the operational impact from future IMS-originated emergency calls.	ATIS-0700015	September 2014	Information Standard		A	O	E	P

Entity	Standard or Document ID	Standard or Document Title	Standard or Document Description	Associated Documents	Latest Revision/ Release Date	Standard or Document Type	Relation to NENA i3 Architecture				
							Client (C)	Access Networks (A)	Origination Networks (O)	ESInets (E)	PSAPs (P)
ATIS	ATIS-0500027 (Fee/Charge)	Recommendations for Establishing Wide Scale Indoor Location Performance	Provides the methodology to characterize wide-scale indoor location accuracy performance by creating regional test beds and extrapolating their test results.		May 2015	Technical Standard		A	O		
ATIS	ATIS-0500028 (Fee/Charge)	Analysis of Unwanted User Service Interactions with NG9-1-1 Capabilities	Illustrates use cases that convey the need for a broader analysis of standardized user service definitions for possible interactions with NG9-1-1 capabilities and identification of which interactions could lead to unwanted behavior.		February 2015	Technical Report		A	O	E	P
ATIS	ATIS-0700015.v003 (Fee/Charge)	ATIS Standard for Implementation of 3GPP Common IMS Emergency Procedures for IMS Origination and ESInet/Legacy Selective Router Termination	Identifies and adapts 3GPP common IMS emergency procedures for applicability in North America to support emergency communications originating from an IMS subscriber.	ATIS-0500026	May 2015	Technical Standard		A	O		

Next Generation 911 (NG911) Standards Identification and Review

Entity	Standard or Document ID	Standard or Document Title	Standard or Document Description	Associated Documents	Latest Revision/ Release Date	Standard or Document Type	Relation to NENA i3 Architecture				
							Client (C)	Access Networks (A)	Origination Networks (O)	ESInets (E)	PSAPs (P)
ATIS	ATIS-1000010.2006(R2011) (Fee/Charge)	Support of Emergency Telecommunications Service ETS in IP Network	Defines the procedures and capabilities required to support ETS within and between IP-based service provider networks.		June 2006	Technical Standard		A	O		
ATIS	ATIS-1000012.2006(R2011) (Fee/Charge)	Signaling System No. 7 (SS7) – SS7 Network and NNI Interconnection Security Requirements and Guidelines	Provides security requirements and guidelines for SS7 network and its network interconnections.		November 2006	Technical Standard		A	O	E	
ATIS	ATIS-1000019.2007(R2012) (Fee/Charge)	Network to Network Interface (NNI) Standard for Signaling and Control Security for Evolving VoP Multimedia Networks	Specifies VoP and multimedia signaling and control plane security requirements for evolving networks.		March 2007	Technical Standard		A	O	E	
ATIS	ATIS-1000023.2013 (Fee/Charge)	ETS Network Element Requirements for A NGN IMS Based Deployments	Defines network element requirements to ensure that ETS is implementable and interoperable in a multi-vendor environment for an NGN IMS-based network deployment; refines the procedures defined in the ETS in IP Networks Phase 1 standard.	ATIS-1000010	August 2013	Technical Standard		A	O		

Entity	Standard or Document ID	Standard or Document Title	Standard or Document Description	Associated Documents	Latest Revision/ Release Date	Standard or Document Type	Relation to NENA i3 Architecture				
							Client (C)	Access Networks (A)	Origination Networks (O)	ESInets (E)	PSAPs (P)
ATIS	ATIS-1000026 .2008(R2013) (Fee/Charge)	Session Border Controller Functions and Requirements	Defines the Session Border Controller (SBC) functions and requirements that reside within a service provider's network.		April 2008	Technical Standard		A	O		
ATIS	ATIS-1000029.2008 (R2013) (Fee/Charge)	Security Requirements for NGN	Provides security requirements for the NGN against security threats, and to mitigate the effects of security attacks.	ATIS-1000034. 2010 (R2015)	November 2008	Technical Standard		A	O	E	
ATIS	ATIS-1000034.2010(R2015) (Fee/Charge)	Next Generation Network (NGN): Security Mechanisms and Procedures	Describes some security mechanisms that can be used to fulfill the requirements described in ATIS-1000029.2008 and specifies the suite of options for each selected mechanism.	ATIS-1000029. 2008	November 2010	Technical Standard		A	O	E	
ATIS	ATIS-1000038 (Fee/Charge)	Technical Parameters for IP Network to Network Interconnection Release 1.0	Specifies the "Interconnection Technical Parameters" that need to be collected and eventually exchanged between two service providers so that they can successfully interconnect IP-based facilities and VoIP services at an NNI.		August 2010	Technical Standard					

Next Generation 911 (NG911) Standards Identification and Review

Entity	Standard or Document ID	Standard or Document Title	Standard or Document Description	Associated Documents	Latest Revision/ Release Date	Standard or Document Type	Relation to NENA i3 Architecture				
							Client (C)	Access Networks (A)	Origination Networks (O)	ESInets (E)	PSAPs (P)
ATIS	ATIS-1000040 (Fee/Charge)	Protocol Suite Profile for IP Network to Network Interconnection Release 1.0	Identifies a set of protocols and specifies their profile so that signaling, media, and network related parameters can be uniformly and consistently utilized across the interconnection interface; supports a service seamlessly across an IP network to network interconnection as identified by the test scenarios defined in ATIS-1000041.	ATIS-1000041	August 2010	Technical Standard					
ATIS	ATIS-1000041 (Fee/Charge)	Test Suites for IP Network to Network Interconnection Release 1.0	Specifies a set of call test scenarios involving SIP and other signaling messages which for various situations may be required to provide an expected reaction to an event or a sequence of events appropriate to the previously signaled message; "expected reaction" is based upon the protocol profile established in the messages that flow across the NNI.	ATIS-1000040	August 2010	Technical Standard					

Next Generation 911 (NG911) Standards Identification and Review

Entity	Standard or Document ID	Standard or Document Title	Standard or Document Description	Associated Documents	Latest Revision/ Release Date	Standard or Document Type	Relation to NENA i3 Architecture				
							Client (C)	Access Networks (A)	Origination Networks (O)	ESInets (E)	PSAPs (P)
ATIS	ATIS-1000049 (Fee/Charge)	End-to-End NGN GETS Call Flows	Describes end-to-end call/session flows for various wireline and wireless access technologies, in addition to the IMS Core Network call/session flows in support of NGN GETS.		August 2011	Technical Standard		A	O		
ATIS	ATIS-1000055.2013 (Fee/Charge)	Emergency Telecommunications Service (ETS): Core Network Security Requirements	Provides a minimum set of common (i.e., independent of network type or technology) and core network security requirements for the protection of ETS in a multi-provider NGN environment.		August 2013	Technical Standard		A	O		
ATIS	ATIS-1000060.2014 (Fee/Charge)	Emergency Telecommunications Service (ETS): Long Term Evolution (LTE) Access Network Security Requirements for National Security/Emergency Preparedness (NS/EP) Next Generation Network (NGN) Priority Services	Provides a minimum set of requirements for the security protection of NS/EP NGN-PS in LTE access networks.		October 2014	Technical Standard		A	O		

Next Generation 911 (NG911) Standards Identification and Review

Entity	Standard or Document ID	Standard or Document Title	Standard or Document Description	Associated Documents	Latest Revision/ Release Date	Standard or Document Type	Relation to NENA i3 Architecture				
							Client (C)	Access Networks (A)	Origination Networks (O)	ESInets (E)	PSAPs (P)
ATIS	ATIS-1000061.2015 (Fee/Charge)	LTE Access Class 14 for National Security and Emergency Preparedness (NS/EP) Communications	Provides operational guidance regarding the assignment and use of the 3GPP LTE specifications for Access Class Barring to support NS/EP NGN-PS.		February 2015	Technical Standard		A	O		
ATIS	ATIS-1000065.2015 (Fee/Charge)	Emergency Telecommunications Service (ETS) Evolved Packet Core (EPC) Network Element Requirements	Specifies ETS requirements for an EPS consisting of the E-UTRAN and EPC for support of NGN GETS voice, NGN GETS video, NGN GETS Guaranteed Bit Rate (GBR) data, and NGN GETS data transport.		February 2015	Technical Standard		A	O		
ATIS	ATIS-1000067.2015 (Fee/Charge)	IP NGN Enhanced Calling Name (eCNAM)	Defines a Calling Name Delivery service in the IP-based NGN; includes a mandatory longer name field and optional additional information about the caller.		August 2015	Technical Standard		A	O		
ATIS	ATIS-1000679.2015 (Fee/Charge)	Interworking between Session Initiation Protocol (SIP) and ISDN User Part	Defines the signaling interworking between the ISDN User Part (ISUP) protocol and SIP in order to support services that can be commonly supported by ISUP and SIP based network domains.		April 2015	Technical Standard		A	O		

Next Generation 911 (NG911) Standards Identification and Review

Entity	Standard or Document ID	Standard or Document Title	Standard or Document Description	Associated Documents	Latest Revision/ Release Date	Standard or Document Type	Client (C)	Access Networks (A)	Origination Networks (O)	ESInets (E)	PSAPs (P)
								Relation to NENA i3 Architecture			
ATIS	ESIF Issue 81	*Applying Common IMS to NG9-1-1 Networks (Stage 2 & 3) Specification*	Defines call processing, transport, or delivery of emergency service calls within the NG9-1-1 network to the appropriate PSAP.		In Development	Technical Issue Documentation		A	O	E	P
ATIS	ESIF Issue 82	*IMS-based Next Generation Emergency Services Network Interconnection*			In Development	Technical Issue Documentation		A	O	E	P
ATIS	ESIF Issue 86	*Technical Report to describe ATIS-0700015 for Public Safety*	Will address technical impact from future IMS-originated emergency calls.	ATIS-0700015	In Development	Technical Issue Documentation		A	O	E	P
ATIS	ESIF Issue 87	*Vertical Axis Measurement Test Methodology*			In Development	Technical Issue Documentation		A	O	E	P
ATIS/TIA	ANSI/J-STD-036-C (Fee/Charge)	*Enhanced Wireless 9-1-1 Phase II*	Defines the messaging required to support information transfer to identify and locate wireless emergency service callers.		June 2011	Technical Standard (Joint TIA/ATIS ANS)		A	O	E	P
ATIS/TIA	ANSI/J-STD-036-C-1 (Fee/Charge)	*Addendum to J-STD-036-C, Enhanced Wireless 9-1-1 Phase II*	Enables an MPC and PDE to assign appropriate COS when delivering data to a PSAP.		October 2013	Technical Standard		A	O	E	P

Entity	Standard or Document ID	Standard or Document Title	Standard or Document Description	Associated Documents	Latest Revision/ Release Date	Standard or Document Type	Relation to NENA i3 Architecture				
							Client (C)	Access Networks (A)	Origination Networks (O)	ESInets (E)	PSAPs (P)
ATIS/TIA	J-STD-110.01.v002 (Fee/Charge)	*Joint ATIS/TIA Implementation Guideline for J-STD-110, Joint ATIS/TIA Native SMS/MMS to 9-1-1 Requirements and Architecture Specification, Release 2*	Addresses CMSP and TCC provider deployment considerations of J-STD-110.v002.	J-STD-110.v002	May 2015	Technical Standard (Joint TIA/ATIS ANS, including J-STD-110.01.v002)		A	O	E	P
ATIS/TIA	J-STD-110.v002 including J-STD-110.01.v002 (Fee/Charge)	*Joint ATIS/TIA Native SMS/MMS to 9-1-1 Requirements and Architecture Specification, Release 2*	Defines the requirements, architecture and procedures for text messaging to 9-1-1 emergency services using native wireless operator SMS capabilities for the existing generation and next generation PSAPs.	J-STD-110.01.v002	May 2015	Technical Standard (Joint TIA/ATIS ANS)		A	O	E	P
ATIS	ESIF-E911-Phase2ReadinessPackage	*Wireless E9-1-1 Phase II Readiness Package*	Supplies PSAPs with a standard method for verifying readiness and providing carriers with complete information to speed implementation.		January 2003	Other		A	O		P
ATIS	NGIIF Issue 27	*Documentation of Operational Procedures for Next Generation Networks Interconnection*			In Development	Technical Issue Documentation		A	O	E	P

Next Generation 911 (NG911) Standards Identification and Review

Entity	Standard or Document ID	Standard or Document Title	Standard or Document Description	Associated Documents	Latest Revision/ Release Date	Standard or Document Type	Relation to NENA i3 Architecture					
							Client (C)	Access Networks (A)	Origination Networks (O)	ESInets (E)	PSAPs (P)	
ATIS	NGIIF Issue 31	Develop New Text Related to Methodologies That Support TDM/IP Caller ID Services, Call Spoofing, Etc.			In Development	Technical Issue Documentation		A	O			
ATIS	PTSC Issue 28	US Standard For IP-IP Network Interconnection - Roadmap Standard			In Development	Technical Issue Documentation		A	O			
ATIS	PTSC Issue 66	NGN Architecture Phase 2			In Development	Technical Issue Documentation		A	O			
ATIS	PTSC Issue 81	ETS Wireline Access Requirements			In Development	Technical Issue Documentation		A	O			
ATIS	PTSC Issue 82	ETS Phase 2 Network Element Requirements			In Development	Technical Issue Documentation		A	O			
ATIS	PTSC Issue 93	NGN Security Planning & Operations Guidelines			In Development	Technical Issue Documentation		A	O	E	P	
ATIS	PTSC Issue 98	ETS Roadmap			In Development	Technical Issue Documentation		A	O	E	P	
ATIS	PTSC Issue 100	Supplement to ATIS-1000010			In Development	Technical Issue Documentation		A	O			
ATIS	PTSC Issue 119	Dynamic Priority for Next Generation Secure Communications			In Development	Technical Issue Documentation		A	O	E		

Next Generation 911 (NG911) Standards Identification and Review

Entity	Standard or Document ID	Standard or Document Title	Standard or Document Description	Associated Documents	Latest Revision/ Release Date	Standard or Document Type	Relation to NENA i3 Architecture — Client (C)	Access Networks (A)	Origination Networks (O)	ESInets (E)	PSAPs (P)
ATIS	WTSC Issue 32	*Support of Public Safety Requirements in LTE Networks*			In Development	Technical Issue Documentation		A	O	E	P
ATIS	WTSC Issue 34	*Automating Location Acquisition for Non-Operator-Managed Over-the-Top VoIP Emergency Services Calls*			In Development	Technical Issue Documentation		A	O	E	
ATIS	WTSC Issue 39	*Public Safety Mission Critical Push to Talk (PTT) Voice Interoperation between Land Mobile Radio (LMR) and Long Term Evolution (LTE) Systems*			In Development	Technical Issue Documentation		A	O	E	
ATIS	WTSC Issue 41	*Commercial Mobile Alerts Service (CMAS) International Roaming*			In Development	Technical Issue Documentation		A	O		
ATIS	WTSC Issue 51	*Location Accuracy Improvements for Emergency Calls*			In Development	Technical Issue Documentation		A	O	E	P

Entity	Standard or Document ID	Standard or Document Title	Standard or Document Description	Associated Documents	Latest Revision/ Release Date	Standard or Document Type	Relation to NENA i3 Architecture				
							Client (C)	Access Networks (A)	Origination Networks (O)	ESInets (E)	PSAPs (P)
BICSI	ANSI/BICSI 002-2014 (Fee/Charge)	Data Center Design and Implementation Best Practices	Provides requirements, guidelines and best practices applicable to any data center, including security, power, cooling, cabling, and other topics.		2014 Edition	Informational Document - Best Practices	C			E	P
BICSI	ANSI/BICSI 003-2014 (Fee/Charge)	Building Information Modeling (BIM) Practices for Information Technology Systems	Provides detailed information about BIM content models and object parameters, setting the recommended levels and guidelines for BIM models.			Best Practices					
BICSI	ANSI/BICSI 005-2013 (Fee/Charge)	Electronic Safety and Security (ESS) System Design and Implementation Best Practices	Provides the requirement and recommendations of a structured cabling infrastructure needed to support security systems.			Best Practices					
BICSI	ANSI/BICSI 006-2015 (Fee/Charge)	Distributed Antenna System (DAS) Design and Implementation Best Practices	Provides requirements and recommendations for the design and installation of a standards-compliant, vendor-neutral DAS that is able to be used for a wide range of applications, environments and locations.		2015 Edition	Informational Document - Best Practices	C				P

Next Generation 911 (NG911) Standards Identification and Review

Entity	Standard or Document ID	Standard or Document Title	Standard or Document Description	Associated Documents	Latest Revision/ Release Date	Standard or Document Type	Relation to NENA i3 Architecture				
							Client (C)	Access Networks (A)	Origination Networks (O)	ESInets (E)	PSAPs (P)
BICSI	ANSI/NECA/BICSI 568-2006 (Fee/Charge)	Standard for Installing Commercial Building Telecommunications Cabling	Describes minimum requirements and procedures for installing the infrastructure for telecommunications.		Version 5 2006	Technical Standard					
BICSI	ANSI/NECA/BICSI 607-2011 (Fee/Charge)	Standard for Telecommunications Bonding and Grounding Planning and Installation Methods for Commercial Buildings	Specifies aspects of planning and installation of telecommunications bonding and grounding systems.		Version 5 2011	Technical Standard					
BICSI	Telecommunications Distribution Methods Manual (TDMM) (Fee/Charge)	Telecommunications Distribution Methods Manual	Is the definitive reference manual for telecommunications and information communications technology infrastructure design.		13th Edition	Informational Document - Best Practices	C				P
BICSI	Information Technology Systems Installation Methods Manual (Fee/Charge)	Information Technology Systems Installation Methods Manual	Provides ICT industry installation practices.		6th Edition	Informational Document - Best Practices					

Entity	Standard or Document ID	Standard or Document Title	Standard or Document Description	Associated Documents	Latest Revision/ Release Date	Standard or Document Type	Client (C)	Access Networks (A)	Origination Networks (O)	ESInets (E)	PSAPs (P)
BICSI	Outside Plant Design Reference Manual (Fee/Charge)	Outside Plant Design Reference Manual	Provides information on traditional infrastructure such as cabling and pathways, but also items not typically found within interior design work, such as right-of-way, permitting and service restoration.		5th Edition	Informational Document - Best Practices					
CableLabs	Invention Disclosure 60399 (Free)	Phased Array Scanner for Fire and Police Applications	Uses a phased array scanner at millimeter wavelengths for public safety applications.		August 21, 2012				O		
CableLabs	Invention Disclosure 60620 (Free)	Terrestrial Wi-Fi based Auto Security and Car Safety Service	Defines an architecture for the cable industry to deliver automotive monitoring services or partner with automobile manufacturers to augment existing deployments to it subscribers.		January 27, 2014				O		
CALEA	Standards for Law Enforcement Agencies (Fee/Charge)	CALEA® Standards for Law Enforcement Agencies	Defines a law enforcement agency's role in administration, operations, and facilities and equipment of communications center under their control.		2010	Operational Standard (Chapter 81 Communications applicable)					P

Next Generation 911 (NG911) Standards Identification and Review

Entity	Standard or Document ID	Standard or Document Title	Standard or Document Description	Associated Documents	Latest Revision/ Release Date	Standard or Document Type	Relation to NENA i3 Architecture				
							Client (C)	Access Networks (A)	Origination Networks (O)	ESInets (E)	PSAPs (P)
CALEA	Standards for Public Safety Communications Agencies (Fee/Charge)	CALEA® Standards for Public Safety Communications Agencies	Provides a management model for agency administration and operations, addressing seven critical areas of communications center operations.		2011	Operational Standard					P
DOC	FIPS-PUB-140-2 (Free)	Security Requirements for Cryptographic Modules	Specifies the security requirements that will be satisfied by a cryptographic module utilized within a security system protecting sensitive but unclassified information.	ISO/IEC 19790:2012	December 2002 (FIPS-PUB-140-3 is currently in development)	Technical Standard		A	O		
DOC	FIPS-PUB-180-4 (Free)	Secure Hash Standards (SHS)	Specifies hash algorithms to detect whether messages have not been altered since they were originally generated.		August 2015	Technical Standard		A	O		
DOC	FIPS-PUB-197 (Free)	Advanced Encryption Standards (AES)	Specifies a FIPS-approved cryptographic algorithm that can be used to protect electronic data; the AES algorithm is a symmetric block cipher that can encrypt and decrypt information.		November 2001	Technical Standard (Data/Design)		A	O		

Entity	Standard or Document ID	Standard or Document Title	Standard or Document Description	Associated Documents	Latest Revision/ Release Date	Standard or Document Type	Relation to NENA i3 Architecture				
							Client (C)	Access Networks (A)	Origination Networks (O)	ESInets (E)	PSAPs (P)
DOC	NIST Cybersecurity Framework (Free)	Framework for Improving Critical Infrastructure Cybersecurity	Consists of standards, guidelines, and practices to promote the protection of critical infrastructure; focuses on using business drivers to guide cybersecurity activities and considering cybersecurity risks as part of the organization's risk management process.		February 12, 2014			A	O	E	P
DOC NIST NSTIC	GTRI NSTIC Trustmark Framework (Free)	Trustmark Framework Technical Specification	Provides normative language that governs the structures that comprise the Trustmark Framework and the rules and policies related to the operational use of these structures.		Version 1.0 October 3, 2014			A	O	E	P
DHS	2014 National Emergency Communications Plan (Free)	2014 National Emergency Communications Plan	Provides recommendations to the emergency response community for maintaining communications during routine operations, as well as disasters and acts of terrorism requiring cross-border, multi-state, and multi-jurisdictional responses.		2014	Technical/ Operational	C				P

Next Generation 911 (NG911) Standards Identification and Review

Entity	Standard or Document ID	Standard or Document Title	Standard or Document Description	Associated Documents	Latest Revision/ Release Date	Standard or Document Type	Relation to NENA i3 Architecture				
							Client (C)	Access Networks (A)	Origination Networks (O)	ESInets (E)	PSAPs (P)
DHS	SAFECOM (Free)	*Emergency Communications Governance Guide for State, Local, Tribal, and Territorial Officials*	Provides recommendations and best practices for public safety officials at all levels of government to establish, assess, and update governance structures that represent all emergency communications capabilities.		September 2015	Operational	C				P
DOJ	CJISD-ITS-DOC-08140-5.4 (Free)	*Criminal Justice Information Services (CJIS) Security Policy*	Contains information security requirements, guidelines, and agreements reflecting the will of law enforcement and criminal justice agencies for protecting the sources, transmission, storage, and generation of Criminal Justice Information (CJI).		October 6, 2015 Version 5.4		C	A		E	P
ETSI	ETSI SR 002 777 (Free)	*Emergency Communications (EMTEL); Test/verification procedure for emergency calls*	Outlines test procedures for emergency calls from individuals (citizens) to authorities.		Version 1.1.1 July 2010	Special Report					

Entity	Standard or Document ID	Standard or Document Title	Standard or Document Description	Associated Documents	Latest Revision/ Release Date	Standard or Document Type	Relation to NENA i3 Architecture				
							Client (C)	Access Networks (A)	Origination Networks (O)	ESInets (E)	PSAPs (P)
ETSI	ETSI TS 101 470 (Free)	Emergency Communications (EMTEL); Total Conversation Access to Emergency Services	Defines conditions for using Total Conversation for emergency services with more media than in the regular voice call providing opportunities to more rapid, reliable and confidence-creating resolution of the emergency service cases.		Version 1.1.1 November 2013	Technical Standard	C				P
ETSI	ETSI TS 102 164 (Free)	Telecommunications and Internet converged Services and Protocols for Advanced Networking (TISPAN); Emergency Location Protocols	Specifies the protocol that is used by the local emergency operator to obtain the location information that is registered on the operator location server.	OMA-TS-MLP-V3_2-20051124-C	Version 1.3.1 September 2006	Technical Standard	C	A	O		P
ETSI	ETSI TR 102 180 (Free)	Emergency Communications (EMTEL); Basis of requirements for communication of individuals with authorities/ organizations in case of distress (Emergency call handling)	Provides the requirements for communication from individuals to authorities and organizations in all types of emergencies.		March 2014	Technical Report	C	A			P

Entity	Standard or Document ID	Standard or Document Title	Standard or Document Description	Associated Documents	Latest Revision/ Release Date	Standard or Document Type	Relation to NENA i3 Architecture				
							Client (C)	Access Networks (A)	Origination Networks (O)	ESInets (E)	PSAPs (P)
ETSI	ETSI TS 102 424 (Free)	Telecommunications and Internet converged Services and Protocols for Advanced Networking (TISPAN); Requirements of the NGN network to support Emergency Communication from Citizen to Authority	Contains the requirements of an NGN to support EMTEL from the citizen to authority.		Version 1.1.1 September 2005	Technical Standard		A	O		
ETSI	ETSI TR 103 170 (Free)	Emergency Communications (EMTEL); Total Conversation Access to Emergency Services	Describes conditions for using Total Conversation for emergency services and makes access of emergency services possible to people with disabilities.		Version 1.1.1 November 2012	Technical Report	C	A	O		P
ETSI	ETSI TS 123 167 (Free)	Universal Mobile Telecommunications System (UMTS); LTE; IP Multimedia Subsystem (IMS) emergency sessions	Defines the stage two service description for emergency services in the IMS, including the elements necessary to support IM emergency services.	3GPP TS 23.167 Version 12.0.0 Release 12	Version 12.0.0 September 2014	Technical Standard (Product-Interface/Design)	C	A			

Entity	Standard or Document ID	Standard or Document Title	Standard or Document Description	Associated Documents	Latest Revision/ Release Date	Standard or Document Type	Relation to NENA i3 Architecture				
							Client (C)	Access Networks (A)	Origination Networks (O)	ESInets (E)	PSAPs (P)
ETSI	ETSI TS 182 009 (Free)	Telecommunications and Internet converged Services and Protocols for Advanced Networking (TISPAN); NGN Architecture to support emergency communication from citizen to authority	Defines the architectural description for emergency services in the IMS, including the elements necessary to support IM emergency services.	3GPP TS 23.09 (Release 7) 3GPP TS 23.167, (Release 7)	Version 2.1.1 October 2008	Technical Standard		A	O		
ETSI	ETSI TS 183 036 (Free)	Telecommunications and Internet converged Services and Protocols for Advanced Networking (TISPAN); ISDN/SIP interworking; Protocol specification	Specifies the stage three Protocol Description of the signaling interworking between ISDN DSS1 protocol and SIP.		Version 3.5.1 August 2012	Technical Specification		A			
ETSI	ETSI TS 187 001 (Free)	Telecommunications and Internet converged Services and Protocols for Advanced Networking (TISPAN); NGN SECurity (SEC); Requirements	Defines the security requirements pertaining to TISPAN NGN Release 3.	TISPAN NGN Release 3	Version 3.7.1 March 2011	Technical Specification		A	O	E	

Next Generation 911 (NG911) Standards Identification and Review

Entity	Standard or Document ID	Standard or Document Title	Standard or Document Description	Associated Documents	Latest Revision/ Release Date	Standard or Document Type	Relation to NENA i3 Architecture				
							Client (C)	Access Networks (A)	Origination Networks (O)	ESInets (E)	PSAPs (P)
ETSI	ETSI TR 187 002 (Free)	Telecommunications and Internet converged Services and Protocols for Advanced Networking (TISPAN); TISPAN NGN Security (NGN_SEC); Threat, Vulnerability and Risk Analysis	Presents the results of the Threat Vulnerability Risk Analysis (TVRA) for the NGN.		Version 3.1.1 April 2011	Technical Report		A	O	E	
ETSI	ETSI TS 187 003 (Free)	Telecommunications and Internet converged Services and Protocols for Advanced Networking (TISPAN); NGN Security; Security Architecture	Defines the security architecture of NGN.		Version 2.3.2 March 2011	Technical Specification		A	O	E	
ETSI	ETSI TS 187 005 (Free)	Telecommunications and Internet converged Services and Protocols for Advanced Networking (TISPAN); NGN Lawful Interception; Stage 1 and Stage 2 definition	Specifies the stage two model for Lawful Interception of TISPAN NGN services.		Version 3.1.1 June 2012	Technical Specification		A	O	E	P

Next Generation 911 (NG911) Standards Identification and Review

Entity	Standard or Document ID	Standard or Document Title	Standard or Document Description	Associated Documents	Latest Revision/ Release Date	Standard or Document Type	Relation to NENA i3 Architecture				
							Client (C)	Access Networks (A)	Origination Networks (O)	ESInets (E)	PSAPs (P)
ETSI	ETSI ES 203 178 (Free)	Functional architecture to support European requirements on emergency caller location determination and transport	Describes the unified functional architecture to support European requirements on emergency caller location determination and transport, in particular for the case where VoIP service provider and one or several network operators - all serving the customer in the establishment of an emergency call - are independent enterprises needing to co-operate to determine the location of the (nomadic) caller.		Version 1.1.1 February 2015	Architectural	C	A	O	E	
ETSI	ETSI ES 282 007 (Free)	Telecommunications and Internet converged Services and Protocols for Advanced Networking (TISPAN); IP Multimedia Subsystem (IMS); Functional architecture	Describes the IMS core component of the TISPAN NGN functional architecture and its relationship to other subsystems and components.	3GPP TS 23.517 (Release 8)	Version 2.1.1 November 2008	Technical Standard (Interface/ Design)	C	A	O		

Next Generation 911 (NG911) Standards Identification and Review

Entity	Standard or Document ID	Standard or Document Title	Standard or Document Description	Associated Documents	Latest Revision/ Release Date	Standard or Document Type	Relation to NENA i3 Architecture				
							Client (C)	Access Networks (A)	Origination Networks (O)	ESInets (E)	PSAPs (P)
FCC TFOPA	TFOPA Working Group 2 (Free)	Task Force on Optimal PSAP Architecture (TFOPA)	Provides recommendations to the Commission regarding actions PSAPs can take to optimize their security, operations, and funding as they migrate to NG9-1-1.		January 29, 2016	Best Practice			O	E	P
FCC CSRIC	CSRIC Best Practices Database (Free)	CSRIC Best Practices	Includes search features by number, text, type and keywords to locate best practices resulting from work performed by CSRIC, NRIC and other related FCC initiatives.		Ongoing	Best Practices	C	A	O	E	P
FCC CSRIC	CSRIC IV Working Group 1 Next Generation 9-1-1 Task 1 Subtask 1 (Free)	Final Report - Investigation into Location Improvements for Interim SMS (Text) to 9-1-1	Reviews approaches to provide enhanced location information and evaluates associated limitations and challenges for SMS text to 9-1-1 services.		June 2014	Report		A	O		P
FCC CSRIC	CSRIC IV Working Group 1 Next Generation 9-1-1 Task 1 Subtask 2 (Free)	Final Report - PSAP Requests for Service for Interim SMS Text-to-9-1-1	Provides recommended best practices for 9-1-1 authorities to utilize when requesting the interim SMS text-to-9-1-1 service.		May 2014	Report		A	O		P

Next Generation 911 (NG911) Standards Identification and Review

Entity	Standard or Document ID	Standard or Document Title	Standard or Document Description	Associated Documents	Latest Revision/ Release Date	Standard or Document Type	Relation to NENA i3 Architecture					
							Client (C)	Access Networks (A)	Origination Networks (O)	ESInets (E)	PSAPs (P)	
FCC CSRIC	CSRIC IV Working Group 1 Next Generation 9-1-1 Task 2 (Free)	*Final Report - Location Accuracy and Testing for Voice-over-LTE Networks*	Provides information on the impact VoLTE implementation will have on carriers' ability to comply with existing wireless E9-1-1 location accuracy levels.	CSRIC III WG3 March 2012	September 2014	Report		A	O			
FCC CSRIC	CSRIC IV Working Group 1 Next Generation 911 Task 3 (Free)	*Final Report - Specification for Indoor Location Accuracy Test Bed*	Provides guidance to the Commission on establishing a permanent entity to design, develop, and manage an ongoing public test bed for indoor location technologies.		June 2014	Report		A	O		P	
FCC CSRIC	CSRIC IV Working Group 4 (Free)	*Cybersecurity Risk Management and Best Practices*	Provides recommendations on voluntary mechanisms to assure communication providers are taking necessary measures to manage cybersecurity risks and implementation guidance to help adapt the voluntary NIST Cybersecurity Framework.	NIST Cyber-security Framework	March 2015	Report		A	O			

Entity	Standard or Document ID	Standard or Document Title	Standard or Document Description	Associated Documents	Latest Revision/ Release Date	Standard or Document Type	Relation to NENA i3 Architecture				
							Client (C)	Access Networks (A)	Origination Networks (O)	ESInets (E)	PSAPs (P)
FGDC	FGDC-STD-016-2011 (Free)	*United States Thoroughfare, Landmark, and Postal Address Data Standard*	Provides a data content, classification, quality, and exchange standard for thoroughfare, landmark and postal addresses, and for address reference systems; provides a complete XML schema description for exchange of address data.		Version 2.0 February 2011	Standard				E	P
ISF	ISF Standard of Good Practice for Information Security (Free Executive Summary, Fee/Charge for Full Standard)	*The ISF Standard of Good Practice for Information Security*	Presents effect practices and tools for IT professionals to personnel to manage information risks; enables compliance with ISO/IEC 27002:2013, COBIT 5 for Information Security and the SANS Top 20 Critical Security Controls.		2014	Technical Guidance		A	O	E	
IEEE	IEEE 802.1AB-2009 (Fee/Charge)	*Station and Media Access Control Connectivity Discovery*	Defines a protocol and a set of managed objects that can be used for discovering the physical topology from adjacent stations in IEEE 802 LANs.		2009	Technical Standard	C	A	O		

Next Generation 911 (NG911) Standards Identification and Review

Entity	Standard or Document ID	Standard or Document Title	Standard or Document Description	Associated Documents	Latest Revision/Release Date	Standard or Document Type	Relation to NENA i3 Architecture — Client (C)	Access Networks (A)	Origination Networks (O)	ESInets (E)	PSAPs (P)
IEEE	IEEE 802.1AC (Fee/Charge)	Media Access Control (MAC) Services Definition	Defines the MAC service found in LANs and MANs, and the Internal Sublayer Service and External Internal Sublayer Service provided within MAC Bridges, in abstract terms of their semantics, primitive actions and events, and the parameters of, interrelationship between, and valid sequences of, these actions and events.		September 14, 2012	Technical Standard		A	O		
IEEE	IEEE 802.11-2012 (Free)	Wireless LAN Medium Access Control (MAC) and Physical Layer (PHY) Specifications	Specifies technical corrections and clarifications to IEEE Standard 802.11 for WLANS as well as enhancements to the existing MAC and PHY functions.		May 2012	Technical Standard (Product/Design)		A	O		
IEEE	IEEE 802.16-2012 (Free)	Air Interface for Broadband Wireless Access Systems	Specifies the air interface, including the MAC and PHY, of combined fixed and mobile point-to-multipoint broadband wireless access (BWA) systems providing multiple services.	ETSI HiperMAN	August 2012	Technical Standard (Product/Design)		A	O		

Next Generation 911 (NG911) Standards Identification and Review

Entity	Standard or Document ID	Standard or Document Title	Standard or Document Description	Associated Documents	Latest Revision/ Release Date	Standard or Document Type	Relation to NENA i3 Architecture Client (C)	Access Networks (A)	Origination Networks (O)	ESInets (E)	PSAPs (P)
IEEE	IEEE 802.23 (Fee/Charge)	*Emergency Services for Internet Protocol (IP) Based Citizen to Authority Communications*	Defines and describes the characteristics associated with voice, data, and multi-media requests across IEEE 802 networks and provides a uniform approach for transferring required data for emergency services requests.		Working Group Disbanded in June 2011	Technical Standard (Product-Interface/Design)		A	O	E	
IEEE	IEEE 1512-2006 (Fee/Charge)	*Common Incident Management Message Sets for Use by Emergency Management Centers*	Addresses the exchange of vital data about public safety and emergency management issues involved in transportation-related events, through common incident management sets.	IEEE 2000; IEEE 1512.1-2003; IEEE 1512.3-2002	2006	Technical Standard					
IEEE	IEEE 1903-2011 (Fee/Charge)	*Functional Architecture of Next Generation Service Overlay Networks*	Specifies a functional architecture for a Next Generation Service Overlay Network, consisting of a set of functional entities), their functions, reference points and information flows to illustrate service interaction and media delivery.		2011	Technical Standard		A	O		
IETF	RFC 2328 (Free)	*OSPF Version 2*	Describes the OSPF protocol implementation.		March 2, 2013	Proposed Technical Standard				E	

Entity	Standard or Document ID	Standard or Document Title	Standard or Document Description	Associated Documents	Latest Revision/ Release Date	Standard or Document Type	Relation to NENA i3 Architecture					
							Client (C)	Access Networks (A)	Origination Networks (O)	ESInets (E)	PSAPs (P)	
IETF	RFC 2474 (Free)	Definition of the Differentiated Services Field (DS Field) in the IPv4 and IPv6 Headers	Defines the fields used by the Differentiated Code Point (DSCP) protocol to provide QoS traffic prioritization in an IP network.		March 2, 2013	Proposed Technical Standard				E		
IETF	RFC 2475 (Free)	An Architecture for Differentiated Services	Describes a protocol that provides QoS in an IP network.		March 2, 2013	Proposed Technical Standard				E		
IETF	RFC 3261 (Free)	SIP: Session Initiation Protocol	Describes SIP, an application-layer control (signaling) protocol for creating, modifying, and terminating sessions (including Internet telephone calls, multimedia distribution, and multimedia conferences) with one or more participants.		December 7, 2015	Proposed Technical Standard (Interface/ Design)		A	O			
IETF	RFC 3262 (Free)	Reliability of Provisional Responses in Session Initiation Protocol (SIP)	Describes an extension to SIP providing reliable provisional response messages; the extension uses the option tag "100rel" and defines the Provisional Response Acknowledgement (PRACK) method.		December 7, 2015	Proposed Technical Standard		A	O	E		

Entity	Standard or Document ID	Standard or Document Title	Standard or Document Description	Associated Documents	Latest Revision/ Release Date	Standard or Document Type	Relation to NENA i3 Architecture Client (C)	Access Networks (A)	Origination Networks (O)	ESInets (E)	PSAPs (P)
IETF	RFC 3263 (Free)	*Session Initiation Protocol (SIP): Locating SIP Servers*	Describes the DNS procedures to resolve SIP URI into the IP address, port, and transport protocol of the next hop to contact.		December 7, 2015	Proposed Technical Standard		A	O	E	
IETF	RFC 3264 (Free)	*An Offer/Answer Model with Session Description Protocol (SDP)*	Describes a mechanism by which two entities can make use of the SDP to arrive at a common view of a multimedia session between them.		March 2, 2013	Proposed Technical Standard		A	O	E	P
IETF	RFC 3265 (Free)	*Session Initiation Protocol (SIP)-Specific Event Notification*	Describes a SIP extension to provide an extensible framework by which SIP nodes can request notification from remote nodes indicating that certain events have occurred.		October 14, 2015	Proposed Technical Standard		A	O	E	P
IETF	RFC 3411 (Free)	*An Architecture for Describing Simple Network Management Protocol (SNMP) Management Frameworks*	Describes an architecture for describing SNMP management frameworks.		October 14, 2015	Proposed Technical Standard				E	P

Entity	Standard or Document ID	Standard or Document Title	Standard or Document Description	Associated Documents	Latest Revision/ Release Date	Standard or Document Type	Relation to NENA i3 Architecture				
							Client (C)	Access Networks (A)	Origination Networks (O)	ESInets (E)	PSAPs (P)
IETF	RFC 3412 (Free)	Message Processing and Dispatching for the Simple Network Management Protocol (SNMP)	Describes the message processing and dispatching for SNMP messages within the SNMP architecture; defines the procedures for dispatching potentially multiple versions of SNMP messages.		October 14, 2015	Proposed Technical Standard				E	P
IETF	RFC 3413 (Free)	Simple Network Management Protocol (SNMP) Applications	Describes five types of SNMP applications that make use of an SNMP engine as described in RFC 3411.		October 14, 2015	Proposed Technical Standard				E	P
IETF	RFC 3414 (Free)	User-based Security Model (USM) for version 3 of the Simple Network Management Protocol (SNMPv3)	Defines the elements of procedure for providing SNMP message level security; includes an MIB for remotely monitoring/ managing the configuration parameters.		October 14, 2015	Proposed Technical Standard				E	P
IETF	RFC 3415 (Free)	View-based Access Control Model (VACM) for the Simple Network Management Protocol (SNMP)	Defines the elements of procedure for controlling access to management information; includes an MIB for remotely managing the configuration parameters.		October 14, 2015	Proposed Technical Standard				E	P

Next Generation 911 (NG911) Standards Identification and Review

Entity	Standard or Document ID	Standard or Document Title	Standard or Document Description	Associated Documents	Latest Revision/ Release Date	Standard or Document Type	Relation to NENA i3 Architecture Client (C)	Access Networks (A)	Origination Networks (O)	ESInets (E)	PSAPs (P)
IETF	RFC 3416 (Free)	Version 2 of the Protocol Operations for the Simple Network Management Protocol (SNMP)	Defines version 2 of the protocol operations for SNMP; defines the syntax and elements of procedure of sending, receiving, and processing SNMP PDUs.		October 14, 2015	Proposed Technical Standard				E	P
IETF	RFC 3417 (Free)	Transport Mappings for the Simple Network Management Protocol (SNMP)	Defines the transport of SNMP messages over various protocols.		October 14, 2015	Proposed Technical Standard				E	P
IETF	RFC 3418 (Free)	Management Information Base (MIB) for the Simple Network Management Protocol (SNMP)	Defines managed objects which describe the behavior of an SNMP entity.		October 14, 2015	Proposed Technical Standard				E	P
IETF	RFC 3550 (Free)	RTP: A Transport Protocol for Real-Time Applications	Describes the Real-time Transport Protocol (RTP), suitable for transmitting real-time information such as voice, video, and other delay-sensitive media.		October 14, 2015	Proposed Technical Standard				E	P
IETF	RFC 3693 (Free)	Geopriv Requirements	Describes the requirements for the Geopriv Location Object and the protocols that use this Location Object.		February 18, 2016	Proposed Technical Standard		A	O	E	P

Entity	Standard or Document ID	Standard or Document Title	Standard or Document Description	Associated Documents	Latest Revision/ Release Date	Standard or Document Type	Relation to NENA i3 Architecture					
							Client (C)	Access Networks (A)	Origination Networks (O)	ESInets (E)	PSAPs (P)	
IETF	RFC 3856 (Free)	*A Presence Event Package for the Session Initiation Protocol (SIP)*	Describes the usage of SIP for subscriptions and notifications of presence.		October 14, 2015	Proposed Technical Standard		A	O			
IETF	RFC 3863 (Free)	*Presence Information Data Format (PIDF)*	Specifies the Common Profile for Presence (CPP) PIDF as a common presence data format.		October 14, 2015	Proposed Technical Standard	C	A	O	E	P	
IETF	RFC 3966 (Free)	*The tel URI for Telephone Numbers*	Specifies the URI scheme "tel," which describes resources identified by telephone numbers.		March 2, 2013	Proposed Technical Standard		A	O			
IETF	RFC 3986 (Free)	*Uniform Resource Identifier (URI): Generic Syntax*	Defines the generic URI syntax and a process for resolving URI references that might be in relative form, along with guidelines and security considerations for the use of URIs on the Internet.		October 14, 2015	Technical Standard		A	O			

Entity	Standard or Document ID	Standard or Document Title	Standard or Document Description	Associated Documents	Latest Revision/ Release Date	Standard or Document Type	Relation to NENA i3 Architecture				
							Client (C)	Access Networks (A)	Origination Networks (O)	ESInets (E)	PSAPs (P)
IETF	RFC 4079 (Free)	*A Presence Architecture for the Distribution of GEOPRIV Location Objects*	Examines some existing IETF work on the concept of presence, shows how presence architectures map onto GEOPRIV architectures, and demonstrates that tools already developed for presence could be reused to simplify the standardization and implementation of GEOPRIV.		March 2, 2013	Technical Information Document		A	O		
IETF	RFC 4119 (Free)	*A Presence-based GEOPRIV Location Object Format*	Describes an object format for carrying geographical information on the Internet.		March 2, 2013	Proposed Technical Standard		A	O		
IETF	RFC 4271 (Free)	*A Border Gateway Protocol 4 (BGP-4)*	Discusses the BGP, which is an inter-Autonomous System routing protocol; provides a set of mechanisms for supporting Classless Inter-Domain Routing.		October 14, 2015	Proposed Technical Standard (Data/Design)		A	O	E	P
IETF	RFC 4975 (Free)	*The Message Session Relay Protocol (MSRP)*	Describes MSRP, a protocol for transmitting a series of related instant messages in the context of a session.		September 2007	Proposed Technical Standard		A	O		P

Entity	Standard or Document ID	Standard or Document Title	Standard or Document Description	Associated Documents	Latest Revision/ Release Date	Standard or Document Type	Relation to NENA i3 Architecture					
							Client (C)	Access Networks (A)	Origination Networks (O)	ESInets (E)	PSAPs (P)	
IETF	RFC 4976 (Free)	Relay Extensions for the Message Sessions Relay Protocol (MSRP)	Introduces the concept of message relay intermediaries to MSRP and describes the extensions necessary to use them.		September 2007	Proposed Technical Standard		A	O		P	
IETF	RFC 5069 (Free)	Security Threats and Requirements for Emergency Call Marking and Mapping	Reviews the security threats associated with the marking of signaling messages to indicate that they are related to an emergency, and with the process of mapping locations to URIs that point to PSAPs.		January 2008	Informational Document		A	O			
IETF	RFC 5139 (Free)	Revised Civic Location Format for Presence Information Data Format Location Object (PIDF-LO)	Defines an XML format for the representation of civic location.		October 14, 2015	Proposed Technical Standard		A	O			
IETF	RFC 5222 (Free)	LoST: A Location-to-Service Translation Protocol	Describes an XML-based protocol for mapping service identifiers and geodetic or civic location information to service contact URIs.		October 14, 2015	Proposed Technical Standard (Interface/ Design)		A	O			

Next Generation 911 (NG911) Standards Identification and Review

Entity	Standard or Document ID	Standard or Document Title	Standard or Document Description	Associated Documents	Latest Revision/ Release Date	Standard or Document Type	Client (C)	Access Networks (A)	Origination Networks (O)	ESInets (E)	PSAPs (P)
IETF	RFC 5223 (Free)	*Discovering Location-to-Service Translation (LoST) Servers Using the Dynamic Configuration Protocol (DHCP)*	Describes how a LoST client can discover a LoST server using DHCP.		October 14, 2015	Proposed Technical Standard		A	O		
IETF	RFC 5246 (Free)	*The Transport Layer Security (TLS) Protocol Version 1.2*	Describes the communications security over the Internet; the protocol allows client/server applications to communicate in a way that is designed to prevent eavesdropping, tampering, or message forgery.		October 2015 (update in progress)	Proposed Technical Standard				E	P
IETF	RFC 5340 (Free)	*OSPF for IPv6*	Describes the modifications to OSPF to support IPv6.		October 14, 2015	Proposed Technical Standard			O	E	
IETF	RFC 5411 (Free)	*A Hitchhiker's Guide to the Session Initiation Protocol (SIP)*	Lists a current snapshot of the specifications under the SIP umbrella, summarizes each, and groups them into categories.		October 14, 2015	Informational Document	C	A	O	E	P

Entity	Standard or Document ID	Standard or Document Title	Standard or Document Description	Associated Documents	Latest Revision/ Release Date	Standard or Document Type	Relation to NENA i3 Architecture				
							Client (C)	Access Networks (A)	Origination Networks (O)	ESInets (E)	PSAPs (P)
IETF	RFC 5491 (Free)	*GEOPRIV Presence Information Data Format Location Object (PIDF-LO) Usage Clarification, Considerations, and Recommendations*	Makes recommendations on how to constrain, represent, and interpret locations in a PIDF-LO; recommends a subset of GML 3.1.1 that is mandatory to implement by applications involved in location-based routing.		October 14, 2015	Proposed Technical Standard		A	O		
IETF	RFC 5582 (Free)	*Location-to-URL Mapping Architecture and Framework*	Describes an architecture for a global, scalable, resilient, and administratively distributed system for mapping geographic location information to URLs, using the LoST protocol.		October 14, 2015	Proposed Technical Standard			O	E	
IETF	RFC 5880 (Free)	*Bidirectional Forwarding Detection (BFD)*	Describes a protocol intended to detect faults in the bidirectional path between two forwarding engines, including interfaces, data link, and the forwarding engines themselves where possible.		October 14, 2015	Proposed Technical Standard			O	E	
IETF	RFC 5881 (Free)	*Bidirectional Forwarding Detection (BFD) for IPv4 and IPv6 (Single Hop)*	Describes the particulars necessary to used BFD in the IPv4 and IPv6 environments.		October 14, 2015	Proposed Technical Standard			O	E	

Entity	Standard or Document ID	Standard or Document Title	Standard or Document Description	Associated Documents	Latest Revision/ Release Date	Standard or Document Type	Relation to NENA i3 Architecture				
							Client (C)	Access Networks (A)	Origination Networks (O)	ESInets (E)	PSAPs (P)
IETF	RFC 5882 (Free)	*Generic Application of Bidirectional Forwarding Detection (BFD)*	Describes the generic application of the BFD protocol.		October 14, 2015	Proposed Technical Standard			O	E	
IETF	RFC 5985 (Free)	*HTTP-Enabled Location Delivery (HELD)*	Defines a Layer 7 Location Configuration Protocol (LCP) and describes the used of HTTP and HTTP/TLS as transports for the L7 LCP.		October 14, 2015	Proposed Technical Standard (Interface/ Design)		A	O		
IETF	RFC 6135 (Free)	*An Alternative Connection Model for the Message Session Relay Protocol (MSRP)*	Defines an alternative connection model MSRP User Agents (UAs); uses the connection-oriented media (COMEDIA) mechanism in order to create the MSRP transport connection.		February 2011	Proposed Technical Standard		A	O		P
IETF	RFC 6155 (Free)	*Use of Device Identity in HTTP-Enabled Location Delivery (HELD)*	Extends the HELD protocol to allow the location request message to carry device identifiers; privacy and security considerations describe the conditions where requests containing identifiers are permitted.		October 14, 2015	Proposed Technical Standard		A	O		
IETF	RFC 6442 (Free)	*Location Conveyance for the Session Initiation Protocol*	Defines an extension to SIP to convey geographic location information from one SIP entity to another SIP entity.		October 14, 2015	Proposed Technical Standard		A	O		

Next Generation 911 (NG911) Standards Identification and Review

Entity	Standard or Document ID	Standard or Document Title	Standard or Document Description	Associated Documents	Latest Revision/ Release Date	Standard or Document Type	Relation to NENA i3 Architecture					
							Client (C)	Access Networks (A)	Origination Networks (O)	ESInets (E)	PSAPs (P)	
IETF	RFC 6446 (Free)	Session Initiation Protocol (SIP) Event Notification Extension for Notification Rate Control	Specifies mechanisms for adjusting the rate of SIP event notifications.		October 14, 2015	Proposed Technical Standard		A	O	E	P	
IETF	RFC 6447 (Free)	Filtering Location Notifications in the Session Initiation Protocol (SIP)	Describes filters that limit asynchronous location notifications to compelling events.		October 14, 2015	Proposed Technical Standard		A	O	E	P	
IETF	RFC 6714 (Free)	Connection Establishment for Media Anchoring (CEMA) for the Message Session Relay Protocol (MSRP)	Defines an MSRP extension, CEMA; support of this extension is optional.		August 2012	Proposed Technical Standard		A	O		P	
IETF	RFC 6739 (Free)	Synchronizing Service Boundaries and <mapping> Elements Based on the Location-to-Service Translation (LoST) Protocol	Defines an XML protocol to exchange mappings between two nodes.		October 14, 2015			A		E		
IETF	RFC 6753 (Free)	A Location Dereference Protocol Using HTTP-Enabled Location Delivery (HELD)	Describes how to use HTTP over TLS as a dereferencing protocol to resolve a reference to a PIDF-LO.		October 2012	Proposed Technical Standard		A	O			

Entity	Standard or Document ID	Standard or Document Title	Standard or Document Description	Associated Documents	Latest Revision/ Release Date	Standard or Document Type	Relation to NENA i3 Architecture					
							Client (C)	Access Networks (A)	Origination Networks (O)	ESInets (E)	PSAPs (P)	
IETF	RFC 6772 (Free)	Geolocation Policy: A Document Format for Expressing Privacy Preferences for Location Information	Defines an authorization policy language for controlling access to location information; extends the Common Policy authorization framework to provide location-specific access control.		January 2013	Proposed Technical Standard	C	A				
IETF	RFC 6739 (Free)	Synchronizing Service Boundaries and <mapping> Elements Based on the Location-to-Service Translation (LoST) Protocol	Defines an XML protocol to exchange mappings between two nodes.		October 2012	Experimental Technical Standard		A	O			
IETF	RFC 6848 (Free)	Specifying Civic Address Extensions in the Presence Information Data Format Location Object (PIDF-LO)	Describes a backward-compatible mechanism for adding civic address elements to the Geopriv civic address format.		January 2013	Proposed Technical Standard		A	O			
IETF	RFC 6881 (Free)	Best Current Practice for Communications Services in Support of Emergency Calling	Describes best current practice on how devices, networks, and services using IETF protocols should use such standards to make emergency calls.		March 2013	Best Current Practice		A	O			

Entity	Standard or Document ID	Standard or Document Title	Standard or Document Description	Associated Documents	Latest Revision/ Release Date	Standard or Document Type	Relation to NENA i3 Architecture				
							Client (C)	Access Networks (A)	Origination Networks (O)	ESInets (E)	PSAPs (P)
IETF	RFC 6915 (Free)	Flow Identity Extension for HTTP-Enabled Location Delivery (HELD)	Specifies an XML schema and a URN sub-namespace for a Flow Identity Extension for HELD.		April 2013	Proposed Technical Standard		A	O		
IETF	RFC 7035 (Free)	Relative Location Representation	Defines an extension to the PIDF-LO for the expression of location information that is defined relative to a reference point.		October 2013	Proposed Technical Standard		A	O		
IETF	RFC 7090 (Free)	Public Safety Answering Point (PSAP) Callback	Discusses shortcomings of the current PSAP call-back mechanisms and illustrates additional scenarios where better-than-normal call treatment behavior would be desirable.		April 2014	Proposed Technical Standard		A	O		
IETF	RFC 7105 (Free)	Using Device-Provided Location-Related Measurements in Location Configuration Protocols	Describes a protocol for a device to provide location-related measurement data to a LIS within a request for location information.		January 2014	Proposed Technical Standard		A	O		

Entity	Standard or Document ID	Standard or Document Title	Standard or Document Description	Associated Documents	Latest Revision/ Release Date	Standard or Document Type	Relation to NENA i3 Architecture				
							Client (C)	Access Networks (A)	Origination Networks (O)	ESInets (E)	PSAPs (P)
IETF	RFC 7163 (Free)	URN for Country-Specific Emergency Services	Updates the registration guidance provided in Section 4.2 of RFC 5031, which allows the registration of service URNs with the 'sos' service type only for emergency services "that are offered widely and in different countries;" updates those instructions to allow such registrations when, at the time of registration, those services are offered in only one country.		March 2014	Proposed Technical Standard		A	O		
IETF	RFC 7199 (Free)	Location Configuration Extensions for Policy Management	Extends the current location configuration protocols to provide hosts with a reference to the rules that are applied to a URI so that the host can view or set these rules.		April 2014	Proposed Technical Standard		A	O		
IETF	RFC 7216 (Free)	Location Information Server (LIS) Discovery Using IP Addresses and Reverse DNS	Describes the configuration challenge of discovering a LIS when a residential gateway is present, requiring a method that is able to work around the obstacle presented by the gateway.		April 2014	Proposed Technical Standard		A	O		

Next Generation 911 (NG911) Standards Identification and Review

Entity	Standard or Document ID	Standard or Document Title	Standard or Document Description	Associated Documents	Latest Revision/ Release Date	Standard or Document Type	Client (C)	Access Networks (A)	Origination Networks (O)	ESInets (E)	PSAPs (P)
								Relation to NENA i3 Architecture			
IETF	RFC 7378 (Free)	*Trustworthy Location*	Describes threats to conveying location, particularly for emergency calls, and describes techniques that improve the reliability and security of location information.		December 2014	Informational Document		A	O	E	P
IETF	RFC 7406 (Free)	*Extensions to the Emergency Services Architecture for Dealing With Unauthenticated and Unauthorized Devices*	Provides a problem statement, introduces terminology and describes an extension for the base IETF emergency services architecture to address scenarios involving situations dealing with unauthenticated and unauthorized devices making emergency calls.		December 2014	Informational Document		A	O		
IETF	RFC 7459 (Free)	*Representation of Uncertainty and Confidence in the Presence Information Data Format Location Object (PIDF-LO)*	Defines key concepts of uncertainty and confidence as they pertain to location information; outlines methods for the manipulation of location estimates that include uncertainty information.		October 14, 2015	Technical Standard		A	O		

Next Generation 911 (NG911) Standards Identification and Review

Entity	Standard or Document ID	Standard or Document Title	Standard or Document Description	Associated Documents	Latest Revision/ Release Date	Standard or Document Type	Relation to NENA i3 Architecture				
							Client (C)	Access Networks (A)	Origination Networks (O)	ESInets (E)	PSAPs (P)
IETF	RFC 7701 (Free)	Multi-party Chat Using the Message Session Relay Protocol (MSRP)	Defines the necessary tools for establishing multi-party chat sessions, or chat rooms, using MSRP.		December 2015	Proposed Technical Standard		A	O		P
IETF	Internet Draft (draft-ietf-ecrit-additional-data-37) (Free)	Additional Data Related to an Emergency Call	Describes data structures and mechanisms to convey information about the call, caller or location to a PSAP.		December 11, 2015 (Draft status expires April 14, 2016)	Draft Technical Standard		A	O		
IETF	Internet Draft (draft-ietf-ecrit-data-only-ea-11) (Free)	Data-Only Emergency Calls	Describes a container for data-only emergency calls (e.g., temperature sensors, burglar alarms, chemical spill sensors) based on the CAP and its transmission using the SIP MESSAGE transaction.		March 1, 2016 (Draft status expires September 2, 2016)	Draft Technical Standard		A	O		
IETF	Internet Draft (draft-ietf-ecrit-held-routing-05) (Free)	A Routing Request Extension for the HELD Protocol	Specifies an extension to the HELD protocol to support location servers returning routing information with the location information id the location request includes a request for the desired routing information.		February 10, 2016 (Draft status expires August 12, 2016)	Draft Technical Standard		A	O		

Entity	Standard or Document ID	Standard or Document Title	Standard or Document Description	Associated Documents	Latest Revision/ Release Date	Standard or Document Type	Relation to NENA i3 Architecture				
							Client (C)	Access Networks (A)	Origination Networks (O)	ESInets (E)	PSAPs (P)
IETF	Internet Draft (draft-ietf-ecrit-similar-location-01) (Free)	A LoST extension to return complete and similar location info	Introduces a new way to provide returned location information in LoST responses that is either of a completed or similar form to the original input civic location, based on whether valid or invalid civic address elements are returned within the findServiceResponse message.		October 19, 2015 (Draft status expires April 21, 2016)	Draft Technical Standard		A	O		
IETF	Internet Draft (draft-ietf-ecrit-car-crash-07) (Free)	Next-Generation Vehicle-Initiated Emergency Calls	Describes how to use IP-based emergency services mechanisms to support the next generation of emergency calls placed by vehicles and conveying vehicle, sensor, and location data related to the crash or incident.		February 19, 2016 (Draft status expires August 22, 2016)	Draft Informational Document	C	A	O		
IETF	Internet Draft (draft-ietf-mmusic-msrp-usage-data-channel-04) (Free)	MSRP over Data Channels	Specifies how MSRP can be instantiated as a data channel sub-protocol.		February 12, 2016 (Draft status expires August 15, 2016)	Draft Technical Standard		A	O		

Next Generation 911 (NG911) Standards Identification and Review

Entity	Standard or Document ID	Standard or Document Title	Standard or Document Description	Associated Documents	Latest Revision/ Release Date	Standard or Document Type	Relation to NENA i3 Architecture				
							Client (C)	Access Networks (A)	Origination Networks (O)	ESInets (E)	PSAPs (P)
IETF	Internet Draft (draft-pd-dispatch-msrp-websocket-10) (Free)	*The WebSocket Protocol as a Transport for the Message Session Relay Protocol (MSRP)*	Specifies a new WebSocket sub-protocol as a reliable transport mechanism between MSRP clients and relays.		February 14, 2016 (Draft status expires August 17, 2016)	Draft Technical Standard		A	O		
IETF	Internet Draft - EXPIRED (draft-yan-siprec-msrp-recording-04) (Free)	*Overview for MSRP Recording based on SIPREC*	Describes how to achieve MSRP channel recording within the mechanism of SIP Recording (SIPREC).		February 27, 2016 EXPIRED	Draft Technical Standard		A	O	E	P
ISO	ISO 19115-1 (Free)	*Geographic information – Metadata – Part 1: Fundamentals*	Defines the schema required for describing geographic information and services by means of metadata; provides information about the identification, the extent, the quality, the spatial and temporal aspects, the content, the spatial reference, the portrayal, distribution, and other properties of digital geographic data and services.		April 1, 2014 First Edition	Technical Standard				E	P

Entity	Standard or Document ID	Standard or Document Title	Standard or Document Description	Associated Documents	Latest Revision/ Release Date	Standard or Document Type	Relation to NENA i3 Architecture				
							Client (C)	Access Networks (A)	Origination Networks (O)	ESInets (E)	PSAPs (P)
ISO	ISO/IEC 20000-1:2011 (Fee/Charge)	Information technology – Service management – Part 1: Service management system requirements	Specifies requirements for the service provider to plan, establish, implement, operate, monitor, review, maintain and improve an SMS; includes the design, transition, delivery and improvement of services to fulfill agreed service requirements.	ISO 20000 Family	April 15, 2011 Edition 2	Operational Standard	C	A	O	E	P
ISO	ISO/IEC 24760-1:2011 (Free)	Information technology – Security techniques – A framework for identity management – Part 1: Terminology and concepts	Defines terms for identity management and specifies core concepts of identity and identity management, and their relationships.		December 15, 2011 Edition 1					E	P
ISO	ISO/IEC 24760-2:2015 (Fee/Charge)	Information technology – Security techniques – A framework for identity management – Part 2: Reference architecture and requirements	Provides guidelines for the implementation of systems for the management of identity information and specifies requirements for the implementation and operation of a framework for identity management.		June 1, 2015 Edition 1					E	P

Entity	Standard or Document ID	Standard or Document Title	Standard or Document Description	Associated Documents	Latest Revision/ Release Date	Standard or Document Type	Relation to NENA i3 Architecture				
							Client (C)	Access Networks (A)	Origination Networks (O)	ESInets (E)	PSAPs (P)
ISO	ISO/IEC 24760-3	*Information technology – Security techniques – A framework for identity management – Part 3: Practice*			Under development					E	P
ISO	ISO/IEC 27000:2016 (Fee/Charge)	*Information technology – Security techniques – Information security management systems – Overview and vocabulary*	Provides an overview of information security management systems (ISMS), and terms and definitions commonly used in the ISMS family of standards.	ISO 27000 Family	February 15, 2016 Edition 4	Operational Standard	C	A	O	E	P
ISO	ISO/IEC 27001:2013 (Fee/Charge)	*Information technology – Security techniques – Information security management systems – Requirements*	Specifies the requirements for establishing, implementing, maintaining and continually improving an ISMS within the context of the organization; includes requirements for the assessment and treatment of information security risks tailored to the needs of the organization.	ISO 27000 Family	October 10, 2013 Edition 2	Operational Requirements	C	A	O	E	P

Entity	Standard or Document ID	Standard or Document Title	Standard or Document Description	Associated Documents	Latest Revision/ Release Date	Standard or Document Type	Relation to NENA i3 Architecture				
							Client (C)	Access Networks (A)	Origination Networks (O)	ESInets (E)	PSAPs (P)
ISO	ISO/IEC 27002:2013 (Fee/Charge)	Information technology – Security techniques – Code of practice for information security controls	Provides guidelines for organizational information security standards and information security management practices including the selection, implementation and management of controls taking into consideration the organization's information security risk environment(s).	ISO 27000 Family	October 1, 2013 Edition 2	Guidelines	C	A	O	E	P
ISO	ISO/IEC 27003:2010 (Fee/Charge)	Information technology – Security techniques – Information security management system implementation guidance	Focuses on the critical aspects needed for successful design and implementation of ISMS; describes the process of ISMS specification and design from inception to the production of implementation plans.	ISO 27000 Family	February 1, 2010 Edition 1	Design Operational	C	A	O	E	P
ISO	ISO/IEC 27004:2009 (Fee/Charge)	Information technology – Security techniques – Information security management – Measurement	Provides guidance on the development and use of measures and measurement in order to assess the effectiveness of an implemented ISMS and controls or groups of controls.	ISO 27000 Family	December 15, 2009 Edition 1	Design Operational	C	A	O	E	P

Next Generation 911 (NG911) Standards Identification and Review

Entity	Standard or Document ID	Standard or Document Title	Standard or Document Description	Associated Documents	Latest Revision/ Release Date	Standard or Document Type	Relation to NENA i3 Architecture				
							Client (C)	Access Networks (A)	Origination Networks (O)	ESInets (E)	PSAPs (P)
ISO	ISO/IEC 27005:2011 (Fee/Charge)	*Information technology – Security techniques – Information security risk management*	Provides guidelines for information security risk management; is designed to assist the satisfactory implementation of information security based on a risk management approach.	ISO 27000 Family	June 1, 2011 Edition 2	Guidelines Operational	C	A	O	E	P
ISO	ISO/IEC 27011:2008 (Fee/Charge)	*Information technology – Security techniques – Information security management guidelines for telecommunications organizations based on ISO/IEC 27002*	Defines guidelines supporting the implementation of information security management in telecommunications organizations.	ISO 27000 Family	December 15, 2008 Edition 1	Guidelines Operational		A	O	E	P
ISO	ISO/IEC 27031:2011 (Fee/Charge)	*Information technology – Security techniques – Guidelines for information and communication technology readiness for business continuity*	Describes the concepts and principles of ICT readiness for business continuity, and provides a framework of methods and processes to identify and specify all aspects for improving an organization's ICT readiness to ensure business continuity.	ISO 27000 Family	March 1, 2011 Edition 1	Guidelines Operational	C	A	O	E	P

Entity	Standard or Document ID	Standard or Document Title	Standard or Document Description	Associated Documents	Latest Revision/ Release Date	Standard or Document Type	Relation to NENA i3 Architecture				
							Client (C)	Access Networks (A)	Origination Networks (O)	ESInets (E)	PSAPs (P)
ISO	ISO/IEC 27033-1:2015 (Fee/Charge)	Information technology – Security techniques – Network security – Part 1: Overview and concepts	Provides an overview of network security and related definitions. It defines and describes the concepts associated with, and provides management guidance on, network security.	ISO 27000 Family	August 15, 2015 Edition 2	Definitions	C	A	O	E	P
ISO	ISO/IEC 27033-2:2012 (Fee/Charge)	Information technology – Security techniques – Network security – Part 2: Guidelines for the design and implementation of network security	Provides guidelines for organizations to plan, design, implement and document network security.	ISO 27000 Family	August 1, 2012 Edition 1	Guidelines	C	A	O	E	P
ISO	ISO/IEC 27033-3:2010 (Fee/Charge)	Information technology – Security techniques – Network security – Part 3: Reference Networking scenarios – Threats, design techniques and control issues	Describes the threats, design techniques and control issues associated with reference network scenarios; provides detailed guidance on the security threats and the security design techniques and controls required to mitigate the associated risks.	ISO 27000 Family	December 15, 2010 Edition 1	Guidelines	C	A	O	E	P

Entity	Standard or Document ID	Standard or Document Title	Standard or Document Description	Associated Documents	Latest Revision/ Release Date	Standard or Document Type	Relation to NENA i3 Architecture				
							Client (C)	Access Networks (A)	Origination Networks (O)	ESInets (E)	PSAPs (P)
ISO	ISO/IEC 27033-4:2014 (Fee/Charge)	Information technology – Security techniques – Network security – Part 4: Securing communications between networks using security gateways	Provides guidance for securing communications between networks using security gateways (firewall, application firewall, intrusion protection system, etc.) in accordance with a documented information security policy of the security gateways.	ISO 27000 Family	March 1, 2014 Edition 1	Guidelines	C	A	O	E	P
ISO	ISO/IEC 27033-5:2013 (Fee/Charge)	Information technology – Security techniques – Network security – Part 5: Securing communications across networks using Virtual Private Networks (VPNs)	Provides guidelines for the selection, implementation, and monitoring of the technical controls necessary to provide network security using VPN connections to interconnect networks and connect remote users to networks.	ISO 27000 Family	August 1, 2013 Edition 1	Guidelines	C	A	O	E	P

Entity	Standard or Document ID	Standard or Document Title	Standard or Document Description	Associated Documents	Latest Revision/ Release Date	Standard or Document Type	Relation to NENA i3 Architecture				
							Client (C)	Access Networks (A)	Origination Networks (O)	ESInets (E)	PSAPs (P)
ISO	ISO/IEC 27035:2011 (Fee/Charge)	Information technology – Security techniques – Information security incident management	Provides a structured and planned approach to: detect, report and assess information security incidents; respond to and manage information security incidents; detect, assess and manage information security vulnerabilities; and continuously improve information security and incident management as a result of managing information security incidents and vulnerabilities.	ISO 27000 Family	September 1, 2011 Edition 1	Operational	C	A	O	E	P
ISO	ISO/IEC 27037:2012 (Fee/Charge)	Information technology – Security techniques – Guidelines for identification, collection, acquisition and preservation of digital evidence	Provides guidelines for specific activities in the handling of digital evidence, which are identification, collection, acquisition and preservation of potential digital evidence that can be of evidential value.	ISO 27000 Family	2012	Guidelines	C	A	O	E	P
ISO	ISO/IEC CD 29003	Information technology – Security techniques – Identity proofing			Under Development	Technical		A	O	E	

Next Generation 911 (NG911) Standards Identification and Review

Entity	Standard or Document ID	Standard or Document Title	Standard or Document Description	Associated Documents	Latest Revision/ Release Date	Standard or Document Type	Relation to NENA i3 Architecture				
							Client (C)	Access Networks (A)	Origination Networks (O)	ESInets (E)	PSAPs (P)
ISO	ISO/IEC 29115:2013 (Fee/Charge)	*Information technology – Security techniques – Entity authentication assurance framework*	Provides a framework for managing entity authentication assurance in a given context.		April 1, 2013 Edition 1	Technical		A	O	E	
ISO	ISO/IEC FDIS 29146	*Information technology – Security techniques – A framework for access management*			Under Development	Technical		A	O	E	
ITU	ITU-T P.800.2 (Free)	*Mean opinion score interpretation and reporting*	Introduces some of the more common types of mean opinion score (MOS) and describes the minimum information that should accompany MOS values to enable them to be correctly interpreted.		May 14, 2013	Technical				E	P
ITU	ITU-T X.509 (Free)	*Information technology – Open Systems Interconnection – The Directory: Public-key and attribute certificate frameworks*	Defines frameworks for public-key certificates and attribute certificates.		October 14, 2012 Edition 7	Technical		A	O	E	P

Next Generation 911 (NG911) Standards Identification and Review

Entity	Standard or Document ID	Standard or Document Title	Standard or Document Description	Associated Documents	Latest Revision/ Release Date	Standard or Document Type	Client (C)	Access Networks (A)	Origination Networks (O)	ESInets (E)	PSAPs (P)
ITU	ITU-T Y.1271 (Free)	Framework(s) on network requirements and capabilities to support emergency telecommunications over evolving circuit-switched and packet-switched networks	Presents an overview of the basic requirements, features, and concepts for emergency telecommunications that evolving networks are capable of providing.		July 18, 2014	Technical		A	O	E	P
ITU	ITU-T Y.2705 (Free)	Minimum security requirements for the interconnection of the Emergency Telecommunications Service (ETS)	Provides minimum security requirements for the inter-network interconnection of ETS, allowing ETS to be supported with the necessary security protection between different national networks with bilateral and/or multilateral agreements in times of disaster and emergencies.		March 1, 2013	Technical		A	O	E	P
ISACA	Voice-over Internet Protocol (VoIP) Audit/Assurance Program (Fee/Charge)	Voice-over Internet Protocol (VoIP) Audit/Assurance Program	Presents effect practices and tools for IT professionals to develop an audit/assurance program for VoIP networks including those that provide special services such as E-911, backup and recovery systems, and interfaces to the PSTN.		2012	Technical Guidance		A	O	E	

Entity	Standard or Document ID	Standard or Document Title	Standard or Document Description	Associated Documents	Latest Revision/ Release Date	Standard or Document Type	Relation to NENA i3 Architecture				
							Client (C)	Access Networks (A)	Origination Networks (O)	ESInets (E)	PSAPs (P)
ISACA	COBIT 5 Assessment Programme (Fee/Charge)	COBIT 5 Assessment Programme	Provides the basis for assessing an enterprise's processes for the governance and management of IT and related services as described in COBIT 5; provides the basis for a robust, dependable assessment approach, details how to undertake an assessment and provides an alternative and less rigorous approach to performing an assessment.	ISO/IEC 15504-2	Version 5	Technical Guidance		A	O	E	
ISACA	Cybersecurity Nexus™ (CSX) (Fee/Charge)		Provides effective practices to cybersecurity professionals to keep organizations and their information more secure.			Technical Guidance		A	O	E	

Next Generation 911 (NG911) Standards Identification and Review

Entity	Standard or Document ID	Standard or Document Title	Standard or Document Description	Associated Documents	Latest Revision/ Release Date	Standard or Document Type	Relation to NENA i3 Architecture				
							Client (C)	Access Networks (A)	Origination Networks (O)	ESInets (E)	PSAPs (P)
NENA	NENA 02-010 v9 (Free)	*NENA Standard Data Formats For 9-1-1 Data Exchange & GIS Mapping*	Sets forth NENA standard formats for ALI-related data exchange between service providers and data base management system providers, a GIS data model, a data dictionary, and formats for data exchange between the ALI database and PSAP controller equipment.		March 28, 2011 Version 9.0	Technical Standard		A	O	E	
NENA	NENA 02-011 v7.1	*NENA Data Standards For Local Exchange Carriers, ALI Service Providers & 9-1-1 Jurisdictions*	Sets forth standards for all service providers involved in providing dial tone to end users; includes database maintenance, quality measurements, interim number portability (INP), local number portability (LNP) and number-pooling standards to be utilized for any 9-1-1 system that provides information for data display.		**May 12, 2012**	**Technical Standard**		A	O	E	P
NENA	NENA 02-014 v1 (Free)	*NENA GIS Data Collection and Maintenance Standards*	Provides necessary guidelines for collecting and maintaining GIS data.		July 17, 2007 Version 1 Update in Progress	Technical Standard		C	A	E	

Entity	Standard or Document ID	Standard or Document Title	Standard or Document Description	Associated Documents	Latest Revision/ Release Date	Standard or Document Type	Relation to NENA i3 Architecture				
							Client (C)	Access Networks (A)	Origination Networks (O)	ESInets (E)	PSAPs (P)
NENA	NENA 02-015 v1 (Free)	*NENA Technical Standard for Reporting and Resolving ANI/ALI Discrepancies and No Records Found for Wireline, Wireless and VoIP Technologies*	Sets forth standards for PSAP jurisdictions, Access Infrastructure Providers, Service Providers and Data Base Management System Providers in reporting and resolving discrepancies that occurred during a 9-1-1 call.		June 6, 2009 Version 1	Technical Standard		A	O	E	P
NENA	NENA 03-509 v1 (Free)	*Femtocell and UMA TID*	Describes the current state of femtocell and UMA deployments with respect to call processing of E9-1-1 calls, and identifies the impacts to PSAPs of receiving and processing calls from femtocells.		January 27, 2011 Version 1	Technical Information Document					P
NENA	NENA 04-005 v1 (Free)	*NENA ALI Query Service Standard*	Defines the NENA XML ALI Query Service (AQS) that specifies new protocols between the PSAP and the Next Generation Emergency Services Network; provides the rationale behind the AQS and how it relates to the current ALI protocol.		November 21, 2006	Technical Standard		A	O	E	P

Next Generation 911 (NG911) Standards Identification and Review

Entity	Standard or Document ID	Standard or Document Title	Standard or Document Description	Associated Documents	Latest Revision/ Release Date	Standard or Document Type	Relation to NENA i3 Architecture				
							Client (C)	Access Networks (A)	Origination Networks (O)	ESInets (E)	PSAPs (P)
NENA	NENA 06-750 v3 (Free)	NENA Technical Requirements Document On Model Legislation E9-1-1 for Multi-Line Telephone Systems	Helps the owners of the MLTS to understand the issues related to identifying the location of users of the system during emergencies.		February 5, 2011 Version 3.0	Requirements Document	C				P
NENA	NENA 08-001 v2 (Free)	NENA Interim VoIP Architecture for Enhanced 9-1-1 Services (i2)	Provides an overview of the VoIP architecture, functional elements, and interfaces, as well as the interface specifications necessary for interconnection with the existing Emergency Services Network infrastructure.		August 11, 2010 Version 2	Technical Standard		A	O	E	
NENA	NENA 08-503 v1 (Free)	VoIP Characteristics Technical Information Document	Is a VoIP primer document to be used by individuals not familiar with VoIP technology.		June 10, 2004 Version 1	Technical Information Document	C	A	O	E	P
NENA	NENA 08-505 v1 (Free)	NENA Recommended Method(s) for Location Determination to Support IP-Based Emergency Services	Describes solutions that meet the proposed requirements for automatically determining the location of IP devices inside a residential broadband network.		December 21, 2006 Version 1	Technical Information Document		A	O	E	

Entity	Standard or Document ID	Standard or Document Title	Standard or Document Description	Associated Documents	Latest Revision/ Release Date	Standard or Document Type	Relation to NENA i3 Architecture				
							Client (C)	Access Networks (A)	Origination Networks (O)	ESInets (E)	PSAPs (P)
NENA	NENA 08-752 v1 (Free)	NENA Technical Requirements Document for Location Information to Support IP-Based Emergency Services	Provides the NENA requirements for providing information to support emergency calling.		December 21, 2006 Version 1	Technical Standard		A	O	E	
NENA	NENA 71-001 v1 (Free)	NENA Standard For NG9-1-1 Additional Data	Describes the use of additional data available with NG9-1-1 (associated with a call, a location, a caller, and a PSAP) that assists in determining the appropriate call routing and handling.		September 17, 2009 Version 1 Update in Progress	Technical Standard (Data/ Design)		A	O	E	P
NENA	NENA 71-501 v1 (Free)	NENA Information Document for Synchronizing Geographic Information System Databases with MSAG & ALI	Provides PSAP management, vendors, and other interested parties the necessary guidelines for synchronizing GIS data with existing 9-1-1 databases.		September 8, 2009 Version 1	Technical and Operational Standard		A	O	E	
NENA	NENA-STA-004.1.1-2014 (Free)	NENA Next Generation 9-1-1 (NG9-1-1) Civic Location Data Exchange Format (CLDXF) Standard	Supports the exchange of U.S. civic location address information about 9-1-1 calls, both within the U.S. and internationally; defines the detailed data elements needed for address data exchange.		March 23, 2014 Version 1	Technical Standard		A	O	E	

Entity	Standard or Document ID	Standard or Document Title	Standard or Document Description	Associated Documents	Latest Revision/ Release Date	Standard or Document Type	Relation to NENA i3 Architecture Client (C)	Access Networks (A)	Origination Networks (O)	ESInets (E)	PSAPs (P)
NENA	NENA-STA-006.1-201X	GIS Data Model for NG9-1-1	Will define the GIS database model that will be used to support NG9-1-1 systems, databases, call routing, call handling, and related processes.		In Progress	Technical Standard					
NENA	NENA-INF-009.1-2014 (Free)	Requirements for a National Forest Guide Information Document	Gathers a set of requirements for a national, authoritative Forest Guide in order to allow an entity to procure the technology and services required from this NG9-1-1 functional element.		August 14, 2014	Information Document	C				P
NENA	NENA/APCO-INF-005 (Free)	NENA/APCO Emergency Incident Data Document (EIDD) Information Document	Provides information on data components that will be used in the future EIDD technical standard to share emergency incident information between and among authorized entities and systems.	Joint NENA/ APCO technical ANSI standard in progress	February 21, 2014	Information Document				E	P
NENA	NENA 70-Draft	Standards for the Provisioning and Maintenance of GIS data to ECRF/LVF	Defines the operational processes and procedures necessary to support the i3 ECRF and LVF; identifies ECRF/LVF performance and implementation tradeoffs for 9-1-1 Authorities' consideration.		In Progress	Technical Standard			O	E	

Next Generation 911 (NG911) Standards Identification and Review

Entity	Standard or Document ID	Standard or Document Title	Standard or Document Description	Associated Documents	Latest Revision/ Release Date	Standard or Document Type	Relation to NENA i3 Architecture				
							Client (C)	Access Networks (A)	Origination Networks (O)	ESInets (E)	PSAPs (P)
NENA	NENA TBD	*Discrepancy, Performance and Audits for NG9-1-1*	Provides requirements for various discrepancy, performance and system audits.		In Progress	Technical Standard				E	P
NENA	NENA-INF-014.1-2015 (Free)	*NENA Information Document for Development of Site/Structure Address Point GIS Data for 9-1-1*	Provides guidelines for the development of a site/structure GIS layer, including sub-address level attribute fields and address point placement.	NENA-STA-006 draft	September 18, 2015	Information Document				E	P
NENA	NENA 71-502 v1 (Free)	*An Overview of Policy Rules for Call Routing and Handling in NG9-1-1*	Provides an overview of what policy rules are, how policy is defined, and the ways that they may be used.		August 24, 2010 Version 1	Technical and Operational Information Document				E	P
NENA	NENA-STA-003.1.1-2014 (Free)	*NENA Standard for NG9-1-1 Policy Routing Rules*	Defines templates to be used when drafting policy rules to address how and where calls are diverted if the target PSAP is unreachable.		December 1, 2014	Technical Standard				E	P
NENA	NENA-INF-011.1-2014 (Free)	*NENA NG9-1-1 Policy Routing Rules Operations Guide*	Assists 9-1-1 Governing Authorities in using Policy Routing Rules during the full life cycle of a NG9-1-1 System.		October 6, 2014	Information Document				E	P

Entity	Standard or Document ID	Standard or Document Title	Standard or Document Description	Associated Documents	Latest Revision/ Release Date	Standard or Document Type	Relation to NENA i3 Architecture					
							Client (C)	Access Networks (A)	Origination Networks (O)	ESInets (E)	PSAPs (P)	
NENA	NENA 75-001 (Free)	NENA Security for Next-Generation 9-1-1 Standard (NG-SEC)	Establishes the minimal guidelines and requirements for the protection of NG9-1-1 assets or elements within a changing business environment.		February 6, 2010 Version 2 In Progress	Technical Standard (Interface/ Design)		A	O	E	P	
NENA	NENA 75-502 v1 (Free)	Next Generation 9-1-1 Security (NG-SEC) Audit Checklist	Provides a summary of the requirements and recommendations detailed in the NG-SEC standard and provides the educated user a method to document an NG-SEC audit.		December 14, 2011 Version 1	Technical Information Document				E	P	
NENA	NENA 08-002 v1 (Free)	NENA Functional and Interface Standards for Next Generation 9-1-1 Version 1.0 (i3)	Describes the ESInet, which is designed as an IP-based inter-network (network of networks) shared by all agencies that may be involved in any emergency; specifies that all calls enter the ESInet using SIP signaling.		December 18, 2007 Version 1	Technical Standard (Interface/ Design)		A	O	E	P	

Entity	Standard or Document ID	Standard or Document Title	Standard or Document Description	Associated Documents	Latest Revision/ Release Date	Standard or Document Type	Relation to NENA i3 Architecture				
							Client (C)	Access Networks (A)	Origination Networks (O)	ESInets (E)	PSAPs (P)
NENA	NENA 08-003 v1 (Free)	Detailed Functional and Interface Specification for the NENA i3 Solution – Stage 3	Builds upon prior NENA publications including i3 requirements and architecture documents and provides a baseline to other NG9-1-1-related specifications.		June 14, 2011 Version 1 Update in Progress – Document to be numbered NENA-STA-010)	Technical Standard		A	O	E	P
NENA	NENA 08-501 v1 (Free)	NENA Technical Information Document on the Network Interface to IP Capable PSAP	Provides technical information to guide manufacturers of network equipment and PSAP CPE in the development of IP-based interfaces between the network and PSAP CPE and to assist E9-1-1 Network Service Providers and PSAPs in implementing such interfaces.		June 15, 2004 Version 1	Technical Information Document	C	A	O	E	P

Entity	Standard or Document ID	Standard or Document Title	Standard or Document Description	Associated Documents	Latest Revision/ Release Date	Standard or Document Type	Relation to NENA i3 Architecture					
							Client (C)	Access Networks (A)	Origination Networks (O)	ESInets (E)	PSAPs (P)	
NENA	NENA 08-506 v1 (Free)	NENA Emergency Services IP Network Design for NG9-1-1 (NID)	Provides network architects, consultants, 9-1-1 entities, and state authorities with the information that will assist them in developing the requirements for and/or designing ESInets today that will be capable of meeting the requirements of an NG9-1-1 system.		December 14, 2011 Update in Progress	Technical Information Document				E		
NENA	NENA 08-751 v1 (Free)	NENA i3 Technical Requirements Document	Specifies the requirements the i3 (Long Term Definition) Standard should meet.		September 28, 2006 Version 1	Technical Standard	C	A	O	E	P	
NENA	NENA 53-507 (Free)	NENA Virtual PSAP Management Operations Information Document (OID)	Guides PSAP staff and policy makers in evaluating and considering the opportunities and challenges presented with NG9-1-1 systems as they relate to personnel and PSAP management.		May 26, 2009 Version 1	Operational Information Document					P	
NENA	NENA 73-501 v1 (Free)	Use Cases & Suggested Requirements for Non-Voice-Centric (NVC) Emergency Services	Identifies suggested requirements for NVC Emergency Service.		January 11, 2011 Version 1	Technical Information Document	C	A	O	E	P	

Entity	Standard or Document ID	Standard or Document Title	Standard or Document Description	Associated Documents	Latest Revision/ Release Date	Standard or Document Type	Client (C)	Access Networks (A)	Origination Networks (O)	ESInets (E)	PSAPs (P)
NENA	NENA-INF-003.1-2013 (Free)	*NENA Potential Points of Demarcation in NG9-1-1 Networks Information Document*	Identifies points of demarcation.		March 21, 2013	Information Document				E	P
NENA	NENA-INF-TBD	*Non-Mobile Wireless and Broadband Connectivity*	Analyzes current wireless home phone, small cell, femtocell and CMRS handsets with Wi-Fi voice capability and makes recommendations for how to provide the most accurate 9-1-1 location information.		In Progress	Information Document	C	A	O	E	P
NENA	NENA-INF-TBD	*NENA Classes of Service*	Assesses current classes of service and position sources and makes recommendations for additional values.		In Progress	Information Document	C	A	O	E	P
NENA	NENA/APCO-REQ-001.1.1-2016 (Free)	*NENA/APCO Next Generation 9-1-1 Public Safety Answering Point Requirements*	Introduces requirements for a NG9-1-1 PSAP that is capable of receiving IP-based signaling and media for delivery of emergency calls conformant to the latest version of the NENA i3 Architecture document.		January 15, 2016	Requirements Document					P

Next Generation 911 (NG911) Standards Identification and Review

Entity	Standard or Document ID	Standard or Document Title	Standard or Document Description	Associated Documents	Latest Revision/ Release Date	Standard or Document Type	Relation to NENA i3 Architecture Client (C)	Access Networks (A)	Origination Networks (O)	ESInets (E)	PSAPs (P)
NENA	NENA 54-750 v1 (Free)	NENA/APCO Human Machine Interface & PSAP Display Requirements (ORD)	Prescribes the requirements for the human machine interface (HMI) display for the NG9-1-1 system.		October 20, 2010 Version 1	Operational Standard					P
NENA	NENA 57-750 v1 (Free)	NG9-1-1 System and PSAP Operational Features and Capabilities Requirements	Contains a list of operational capabilities or features that are expected to be supported in a standards-based NG9-1-1 system.		June 14, 2011 Version 1	Operational Standard					P
NENA	NENA 04-004 v1 (Free)	NENA Recommended Generic Standards for E9-1-1 PSAP Intelligent Workstations	Defines IWS equipment requirements for users, manufacturers and providers of E9-1-1 CPE.		June 16, 2000 Version 1	Technical Standard					P
NENA	NENA-INF-007.1-2013 (Free)	NENA Information Document for Handling Text-to-9-1-1 in the PSAP	Provides a guideline for PSAPs with recommendations for emergency calling to 9-1-1 using text messaging.		October 9, 2013	Information Document					P
NENA	NENA-INF-012.2-2015 (Free)	NENA Inter-Agency Agreements Model Recommendations Information Document	Provides a model recommendation for the development of mutual aid agreements and MOUs between PSAPs and affiliated or support organizations.		January 8, 2015	Information Document				E	P

Next Generation 911 (NG911) Standards Identification and Review

Entity	Standard or Document ID	Standard or Document Title	Standard or Document Description	Associated Documents	Latest Revision/ Release Date	Standard or Document Type	Relation to NENA i3 Architecture				
							Client (C)	Access Networks (A)	Origination Networks (O)	ESInets (E)	PSAPs (P)
NENA	NENA-REF-002.2-2014 (Free)	PSAP Interim Text-to-9-1-1 Support Documents	Provides support information and education materials for PSAPs planning on moving forward with the interim solution for Text-to-9-1-1.	NENA-REF-003.1-2015:	December 2014	Information Documents					P
NENA	NENA-REF-003.1-2015 (Free)	NENA Text-to-9-1-1 Public Education	Provides guidance when reaching out to local decision makers to educate them on NG9-1-1.	NENA-REF-002.2-2014	March 31, 2015	Information Documents					P
NENA	SMS Text-to-9-1-1 Resources for PSAPs & 9-1-1 Authorities (Free)		Provides public education guidelines, logos and planning strategies.								P
NENA		9-1-1 Authorities Guide to NG9-1-1			In Progress	Operational Standard				E	P
NENA	NG9-1-1 Public Education Plan for Elected Officials and Decision Makers (Free)	Recommended NG9-1-1 Public Education Plan for Elected Officials and Decision Makers	Provides guidance when reaching out to local decision makers to educate them on NG9-1-1 basics and the need to address funding, legislative and regulatory issues to enable the transition to NG9-1-1.		September 24, 2013	Information Document					P
NENA	NENA-STA-008.2-2014 (Free)	NENA Registry System Standard	Describes how registries are created and maintained in NENA.		October 6, 2014 Version 1	Joint Technical (Data) and Operational Standard				E	P

Next Generation 911 (NG911) Standards Identification and Review

Entity	Standard or Document ID	Standard or Document Title	Standard or Document Description	Associated Documents	Latest Revision/ Release Date	Standard or Document Type	Relation to NENA i3 Architecture				
							Client (C)	Access Networks (A)	Origination Networks (O)	ESInets (E)	PSAPs (P)
NENA	NENA-INF-TBD	*Monitoring and Managing NG9-1-1*	Will address specific operational topics and procedures associated with the transition to monitoring and managing NG9-1-1 software functions and infrastructure.		In Progress	Operational Information Document				E	P
NENA	NENA-INF-008.2-2013 (Free)	*NENA NG9-1-1 Transition Plan Considerations Information Document*	Focuses on the aspect of transitioning data from the legacy environment to the NG9-1-1 environment.		November 20, 2013 Version 2 Version 3 in progress will address operational impacts	Information Document		A	O	E	P
NENA	NENA-INF-006.1-2014 (Free)	*NENA NG9-1-1 Planning Guidelines Information Document*	Provides guidance to help 9-1-1 Authorities create a smooth, timely and efficient transition plan to accomplish implementation of NG9-1-1.		January 8, 2014	Information Document					P
NENA	Next Generation 9-1-1 Transition Policy Implementation Handbook (Free)	*Next Generation 9-1-1 Transition Policy Implementation Handbook*	Provides guidance for 9-1-1 leaders and government officials responsible for ensuring that federal, state and local 9-1-1 laws and regulations effectively enable the implementation of NG9-1-1 systems.		March 2010	Best Practice					P

Next Generation 911 (NG911) Standards Identification and Review

Entity	Standard or Document ID	Standard or Document Title	Standard or Document Description	Associated Documents	Latest Revision/ Release Date	Standard or Document Type	Relation to NENA i3 Architecture				
							Client (C)	Access Networks (A)	Origination Networks (O)	ESInets (E)	PSAPs (P)
NENA	NENA ADM-000.18-2014 (Free)	*NENA Master Glossary of 9-1-1 Terminology*	Guide for readers of NENA publications and a tool for members of the NENA committees that prepare them. It defines the terms, acronyms, and definitions associated with the 9-1-1 industry. Intended users of this document are any person needing NENA's definition/description of a 9-1-1 related term.		July 29, 2014	Information Document	C	A	O	E	P
NFPA	NFPA 70 (Fee/Charge) (Free on-line read-only access)	*National Electrical Code® (NEC)*	Addresses the installation of electrical conductors, equipment, and raceways; signaling and communications conductors, equipment, and raceways; and optical fiber cables and raceways in commercial, residential, and industrial occupancies.		2014 Edition	Technical Standard		A	O	E	P

Entity	Standard or Document ID	Standard or Document Title	Standard or Document Description	Associated Documents	Latest Revision/ Release Date	Standard or Document Type	Relation to NENA i3 Architecture				
							Client (C)	Access Networks (A)	Origination Networks (O)	ESInets (E)	PSAPs (P)
NFPA	NFPA 72 (Fee/Charge) (Free on-line read-only access)	National Fire Alarm and Signaling Code	Provides the latest safety provisions to meet society's changing fire detection, signaling, and emergency communications demands; includes requirements for mass notification systems used for weather emergencies; terrorist events; biological, chemical, and nuclear emergencies; and other threats.		2016 Edition	Technical Standard		A	O	E	P
NFPA	NFPA 76 (Fee/Charge) (Free on-line read-only access)	Standard for the Fire Protection of Telecommunications Facilities	Provides requirements for fire protection of telecommunications facilities providing telephone, data, internet transmission, wireless, and video services to the public as well as life safety for the occupants plus protection of equipment and service continuity.		2016 Edition			A	O	E	
NFPA	NFPA 1061 (Fee/Charge) (Free on-line read-only access)	Professional Qualifications for Public Safety Telecommunications Personnel	Identifies the minimum job performance requirements for public safety telecommunicators.		2014 Edition	Operational Standard				E	P

Next Generation 911 (NG911) Standards Identification and Review

Entity	Standard or Document ID	Standard or Document Title	Standard or Document Description	Associated Documents	Latest Revision/ Release Date	Standard or Document Type	Relation to NENA i3 Architecture					
							Client (C)	Access Networks (A)	Origination Networks (O)	ESInets (E)	PSAPs (P)	
NFPA	NFPA 1201 (Fee/Charge) (Free on-line read-only access)	Standard for Providing Fire and Emergency Services to the Public	Contains requirements on the structure and operations of fire emergency service organizations to help protect lives, property, critical infrastructure, and the environment from the effects of hazards.		2015 Edition	Technical Standard					P	
NFPA	NFPA 1221 (Fee/Charge) (Free on-line read-only access)	Standard for the Installation, Maintenance, and Use of Emergency Services Communications Systems	Defines and describes the installation, performance, operation, and maintenance of public emergency services communications systems and facilities.		2016 Edition	Technical Standard				E	P	
NFPA	NFPA 1600 (Fee/Charge) (Free on-line read-only access)	Standard on Disaster/Emergency Management and Business Continuity/ Continuity of Operations Programs	Covers the development, implementation, assessment, and maintenance of programs for prevention, mitigation, preparedness, response, continuity, and recovery.		2016 Edition	Operational Standard						

Entity	Standard or Document ID	Standard or Document Title	Standard or Document Description	Associated Documents	Latest Revision/ Release Date	Standard or Document Type	Relation to NENA i3 Architecture				
							Client (C)	Access Networks (A)	Origination Networks (O)	ESInets (E)	PSAPs (P)
NIEM	NIEM 3.1	National Information Exchange Model	Designed to develop, disseminate and support enterprise-wide information exchange standards and processes that can enable jurisdictions to effectively share critical information in emergency situations, as well as support the day-to-day operations of agencies throughout the U.S.		May 29, 2015 Version 3.1	Technical Architecture				E	
NERC	CIP-002-3 (Free)	Cyber Security — Critical Cyber Asset Identification	Provides a cybersecurity framework for the identification and protection of Critical Cyber Assets to support reliable operation of the Bulk Electric System (BES).		December 2009 Version 3 (Expires March 2016)	Operational Standard		A	O	E	P
NERC	CIP-002-5.1 (Free)	Cyber Security — BES Cyber System Categorization	Identifies and categorizes BES Cyber Systems and their associated BES Cyber Assets for the application of cyber security requirements commensurate with the adverse impact that loss, compromise, or misuse of those BES Cyber Systems could have on the reliable operation of the BES.		April 2016 Version 5	Operational Standard		A	O	E	P

Entity	Standard or Document ID	Standard or Document Title	Standard or Document Description	Associated Documents	Latest Revision/ Release Date	Standard or Document Type	Relation to NENA i3 Architecture					
							Client (C)	Access Networks (A)	Origination Networks (O)	ESInets (E)	PSAPs (P)	
NERC	CIP-003-3 (Free)	*Cyber Security — Security Management Controls*	Requires that Responsible Entities have minimum security management controls in place to protect Critical Cyber Assets.		November 2013 Version 3 Expires March 2016	Operational Standard		A	O	E	P	
NERC	CIP-003-5 (Free)	*Cyber Security — Security Management Controls*	Specifies consistent and sustainable security management controls that establish responsibility and accountability to protect BES Cyber Systems against compromise that could lead to misoperation or instability in the BES.		April 2016 Version 5	Pending Standard		A	O	E	P	
NERC	CIP-004-3a (Free)	*Cyber Security — Personnel & Training*	Requires that personnel having authorized cyber or authorized unescorted physical access to Critical Cyber Assets, including contractors and service vendors, have an appropriate level of personnel risk assessment, training, and security awareness.		May 2012 Version 3a Expires March 2016	Operational Standard		A	O	E	P	

Entity	Standard or Document ID	Standard or Document Title	Standard or Document Description	Associated Documents	Latest Revision/ Release Date	Standard or Document Type	Relation to NENA i3 Architecture				
							Client (C)	Access Networks (A)	Origination Networks (O)	ESInets (E)	PSAPs (P)
NERC	CIP-004-5.1 (Free)	Cyber Security — Personnel & Training	Minimizes the risk against compromise that could lead to misoperation or instability in the BES from individuals accessing BES Cyber Systems by requiring an appropriate level of personnel risk assessment, training, and security awareness in support of protecting BES Cyber Systems.		April 2016 Version 5.1	Operational Standard		A	O	E	P
NERC	CIP-005-3a (Free)	Cyber Security — Electronic Security Perimeter(s)	Requires the identification and protection of the Electronic Security Perimeter(s) inside which all Critical Cyber Assets reside, as well as all access points on the perimeter.		May 2011 Version 3a Expires March 2016	Operational Standard		A	O	E	P
NERC	CIP-005-5 (Free)	Cyber Security — Electronic Security Perimeter(s)	Manages electronic access to BES Cyber Systems by specifying a controlled Electronic Security Perimeter in support of protecting BES Cyber Systems against compromise that could lead to misoperation or instability in the BES.		April 2016 Version 5	Pending Standard		A	O	E	P

Entity	Standard or Document ID	Standard or Document Title	Standard or Document Description	Associated Documents	Latest Revision/ Release Date	Standard or Document Type	Relation to NENA i3 Architecture				
							Client (C)	Access Networks (A)	Origination Networks (O)	ESInets (E)	PSAPs (P)
NERC	CIP-006-3d (Free)	Cyber Security — Physical Security of Critical Cyber Assets	Ensures the implementation of a physical security program for the protection of Critical Cyber Assets.		March 2013 Version 3d/4d Expires March 2016	Operational Standard		A	O	E	P
NERC	CIP-006-5 (Free)	Cyber Security — Physical Security of BES Cyber Systems	Manages physical access to BES Cyber Systems by specifying a physical security plan in support of protecting BES Cyber Systems against compromise that could lead to misoperation or instability in the BES.		April 2016 Version 5	Pending Standard		A	O	E	P
NERC	CIP-007-3a (Free)	Cyber Security — Systems Security Management	Requires Responsible Entities to define methods, processes, and procedures for securing those systems determined to be Critical Cyber Assets, as well as the other (non-critical) Cyber Assets within the Electronic Security Perimeter(s).		November 2013 Version 3a Expires March 2016	Operational Standard		A	O	E	P

Next Generation 911 (NG911) Standards Identification and Review

Entity	Standard or Document ID	Standard or Document Title	Standard or Document Description	Associated Documents	Latest Revision/ Release Date	Standard or Document Type	Relation to NENA i3 Architecture				
							Client (C)	Access Networks (A)	Origination Networks (O)	ESInets (E)	PSAPs (P)
NERC	CIP-007-5 (Free)	Cyber Security — System Security Management	Manages system security by specifying select technical, operational, and procedural requirements in support of protecting BES Cyber Systems against compromise that could lead to misoperation or instability in the BES.		April 2016 Version 5	Pending Standard		A	O	E	P
NERC	CIP-008-3 (Free)	Cyber Security — Incident Reporting and Response Planning	Ensures the identification, classification, response, and reporting of Cyber Security Incidents related to Critical Cyber Assets.		December 2009 Version 3 Expires March 2016	Operational Standard		A	O	E	P
NERC	CIP-008-5 (Free)	Cyber Security — Incident Reporting and Response Planning	Mitigates the risk to the reliable operation of the BES as the result of a Cyber Security Incident by specifying incident response requirements.		April 2016 Version 5	Pending Standard		A	O	E	P
NERC	CIP-009-3 (Free)	Cyber Security — Recovery Plans for Critical Cyber Assets	Ensures that recovery plan(s) are put in place for Critical Cyber Assets and that these plans follow established business continuity and disaster recovery techniques and practices.		December 2009 Version 3 Expires March 2016	Operational Standard		A	O	E	P

Entity	Standard or Document ID	Standard or Document Title	Standard or Document Description	Associated Documents	Latest Revision/ Release Date	Standard or Document Type	Relation to NENA i3 Architecture				
							Client (C)	Access Networks (A)	Origination Networks (O)	ESInets (E)	PSAPs (P)
NERC	CIP-009-5 (Free)	*Cyber Security — Recovery Plans for BES Cyber Systems*	Recovers reliability functions performed by BES Cyber Systems by specifying recovery plan requirements in support of the continued stability, operability, and reliability of the BES.		April 2016 Version 5	Pending Standard		A	O	E	P
NERC	CIP-010-1 (Free)	*Cyber Security — Configuration Change Management and Vulnerability Assessments*	Prevents and detects unauthorized changes to BES Cyber Systems by specifying configuration change management and vulnerability assessment requirements in support of protecting BES Cyber Systems from compromise that could lead to misoperation or instability in the BES.		April 2016 Version 1	Pending Standard		A	O	E	P
OASIS	OASIS CAP v1.2 (Free)	*Common Alerting Protocol*	Defines and describes CAP, which provides an open, non-proprietary digital message format for all types of alerts and notifications.		July 2010 Version 1.2	Technical Standard	C	A	O	E	P

Entity	Standard or Document ID	Standard or Document Title	Standard or Document Description	Associated Documents	Latest Revision/ Release Date	Standard or Document Type	Relation to NENA i3 Architecture				
							Client (C)	Access Networks (A)	Origination Networks (O)	ESInets (E)	PSAPs (P)
OASIS	OASIS EDXL-DE v1.0 (Free)	Emergency Data Exchange Language (EDXL) Distribution Element, v. 1.0	Describes a standard message distribution framework for data sharing among emergency information systems using the XML-based EDXL.		May 1, 2006 Version 1.0	Technical Standard				E	P
OASIS	OASIS EDXL-HAVE (Free)	Emergency Data Exchange Language (EDXL) Hospital AVailability Exchange Version 1.0	Specifies an XML document format that allows the communication of the status of a hospital, its services and resources.		December 2009 Version 1.0 Update in Progress	Technical Standard				E	P
OASIS	OASIS EDXL-RM (Free)	Emergency Data Exchange Language Resource Messaging (EDXL-RM) 1.0	Describes a suite of standard messages for data sharing among emergency and other information systems that deal in requesting and providing emergency equipment, supplies, people and teams.		December 2009 Version 1.0	Technical Standard				E	P

Entity	Standard or Document ID	Standard or Document Title	Standard or Document Description	Associated Documents	Latest Revision/ Release Date	Standard or Document Type	Client (C)	Access Networks (A)	Origination Networks (O)	ESInets (E)	PSAPs (P)
OASIS	OASIS EDXL-SitRep v1.0 (Free)	*Emergency Data Exchange Language Situation Reporting (EDXL-SitRep) Version 1.0*	Describes a set of standard reports and elements that can be used for data sharing among emergency information systems, and that provide incident information for situation awareness on which incident command can base decisions.		April 11, 2013 Version 1.0	Technical Standard				E	P
OASIS	OASIS EDXL-TEC (Free)	*Emergency Data Exchange Language (EDXL) Tracking of Emergency Clients (TEC) Client Registry Exchange Version 1.0*	Provides a standard messaging format for the creation and exchange of client records in and among publicly-accessible registries to assist in tracking and repatriation of displaced individuals during emergencies, disasters, and routine day-to-day incidents.		June 13, 2014 Version 1.0	Technical Standard				E	P
OASIS	OASIS EDXL-TEP (Free)	*Emergency Data Exchange Language (EDXL) Tracking of Emergency Patients (TEP) Version 1.1*	Provides XML messaging standard for exchange of emergency patient and tracking information during patient encounter through admission or release.		December 2013 Version 1.0 Version 1.1 in draft	Technical Standard				E	P

Entity	Standard or Document ID	Standard or Document Title	Standard or Document Description	Associated Documents	Latest Revision/ Release Date	Standard or Document Type	Relation to NENA i3 Architecture				
							Client (C)	Access Networks (A)	Origination Networks (O)	ESInets (E)	PSAPs (P)
OGC	OGC 04-094 Free	Web Feature Service Implementation Standard	Defines interfaces for data access and manipulation operations on geographic features using HTTP as the distributed computing platform.		May 3, 2005 Version 1.1.0	Technical Standard			O	E	
OGC	OGC 06-042 Free	OpenGIS® Web Map Server Implementation Specification	Specifies the behavior of a service that produces spatially referenced maps dynamically from geographic information; specifies operations to retrieve a description of the maps offered by a server to retrieve a map, and to query a server about features displayed on a map.		March 15, 2006 Version 1.3.0	Technical Standard			O	E	
OGC	OGC 07-006r1 Free	OpenGIS® Catalogue Services Specification	Specifies the interfaces, bindings, and a framework for defining application profiles required to publish and access digital catalogues of metadata for geospatial data, services, and related resource information.		February 23, 2007 Version 2.02	Technical Standard			O	E	P

Next Generation 911 (NG911) Standards Identification and Review

Entity	Standard or Document ID	Standard or Document Title	Standard or Document Description	Associated Documents	Latest Revision/ Release Date	Standard or Document Type	Relation to NENA i3 Architecture Client (C)	Access Networks (A)	Origination Networks (O)	ESInets (E)	PSAPs (P)
OGC	OGC 07-074 (Free)	OpenGIS® Location Services (OpenLS): Core Services	Defines OpenLS: Core Services, Parts 1-5, which consists of the composite set of basic services comprising the OpenLS Platform.		September 9, 2008 Version 1.2	Technical Standard			O	E	
OGC	OGC 09-025r2 (Free)	OGC® Web Feature Service 2.0 Interface Standard	Specifies discovery operations, query operations, locking operations, transaction operations and operations to manage stored, parameterized query expressions.		Version 2.0.2 July 2014	Technical Standard			O	E	
OGC	OGC 09-083r3 Free	GeoAPI 3.0 Implementation Standard	Defines application programming interface (API) which can be used for the manipulation of geographic information.		April 25, 2011 Version 3.0.0	Technical Standard			O	E	P
OGC	OGC 10-129r1 Free	OGC® Geography Markup Language (GML) – Extended schemas and encoding rules	Defines the XML schema syntax, mechanisms and conventions that provide an open, vendor-neutral framework for the description of geospatial application schemas for the transport and storage of geographic information in XML.		February 7, 2012 Version 3.3.0	Technical Standard			O	E	

Next Generation 911 (NG911) Standards Identification and Review

Entity	Standard or Document ID	Standard or Document Title	Standard or Document Description	Associated Documents	Latest Revision/ Release Date	Standard or Document Type	Relation to NENA i3 Architecture				
							Client (C)	Access Networks (A)	Origination Networks (O)	ESInets (E)	PSAPs (P)
OGC	OGC 11-030r1 Free	OGC®: Open GeoSMS Standard – Core	Defines an encoding for location enabling a text message to be communicated using SMS.		January 19, 2012 Version 1.0	Technical Standard			O	E	
OGC	OGC 12-019 (Free)	OGC City Geography Markup Language (CityGML) Encoding Standard	Is an open data model and XML-based format for the storage and exchange of virtual 3D city models.		March 9, 2012 Version 2.0.0	Technical Standard			O	E	P
OGC	OGC KML 2.3 (Free)	OGC KML 2.3	Defines three conformance classes (levels) for KML resources, indicating the relative importance or priority of a particular set of constraints; the highest level (CL3) indicates full conformance.		August 4, 2015 Version 1.0	Technical Standard			O	E	P
OMA	OMA-ERP-SUPL-V3_0_2-20110920-C (Free)	OMA Secure User Plane Location V3.0	Outlines the enabler release definition for SUPL Enabler and the respective conformance requirements for clients and servers claiming compliance to it as defined by OMA across the specification baseline.		Candidate September 20, 2011 Version 3.0.2	Technical Standard		A	O		

Next Generation 911 (NG911) Standards Identification and Review

Entity	Standard or Document ID	Standard or Document Title	Standard or Document Description	Associated Documents	Latest Revision/ Release Date	Standard or Document Type	Client (C)	Access Networks (A)	Origination Networks (O)	ESInets (E)	PSAPs (P)
							Relation to NENA i3 Architecture				
OMA	OMA-ERELD-LPPe-V2_0-20141202-C (Free)	OMA LPP Extensions (LPPe) v2.0	Outlines the enabler release definition for LPPe Enabler and the respective conformance requirements for clients and servers claiming compliance to it as defined by OMA across the specification baseline.		Candidate December 2014 Version 2.0	Technical Standard		A	O		
OMA	OMA-ERP-MLP-V3_1-20110920-A (Free)	OMA Mobile Location Protocol V3.1	Identifies the MLP, an application-level protocol for getting the position of mobile stations independent of underlying network technology.		Approved September 20, 2011 Version 3.1	Technical Standard		A	O		
OMA	OMA-ERELD-LOCSIP-V1_0-201201717-A (Free)	OMA Location in SIP/IP Core V1.0	Provides mechanisms to expose location information to application servers connected to a SIP/IP core network.		Approved Version 1.0 January 17, 2012	Technical Standard		A	O		
OMA	OMA SEC_CF 1.1	OMA Application Layer Security Common Functions V1.1	Supports OMA Push services, enablers over SIP and UDP protocols, delegated authentication for Web services, and DTLS, GBA Push, and IPSec profiles.		Candidate Version 1.1 July 31, 2012	Technical Standard		A	O		

Entity	Standard or Document ID	Standard or Document Title	Standard or Document Description	Associated Documents	Latest Revision/ Release Date	Standard or Document Type	Relation to NENA i3 Architecture					
							Client (C)	Access Networks (A)	Origination Networks (O)	ESInets (E)	PSAPs (P)	
SCTE	ANSI/SCTE 24-1 2009 (Free)	IPCablecom 1.0 Part 1: Architecture Framework for the Delivery of Time Critical Services Over Cable Television Networks Using Cable Modems	Provides the architectural framework that will enable cable television operators to provide time-critical services over their networks that have been enhanced to support cable modems.	IPCable-com Series	2009 (In Review)	Technical			O			
SCTE	ANSI/SCTE 24-2 2009 (Free)	IPCablecom 1.0 Part 2: Audio Codec Requirements for the Provision of Bi-directional Audio Service Over Cable Television Networks Using Cable Modems	This standard specifies the audio (voice) codes that are to be used in the provisioning of bi-directional audio services over cable television distribution networks using IP technology.	IPCable-com Series	2009 (In Review)	Technical			O			
SCTE	ANSI/SCTE 24-3 2009 (Free)	IPCablecom Part 3: Network Call Signaling Protocol for the Delivery of Time-Critical Services over Cable Television Using Data Modems	Describes a profile of the Media Gateway Control Protocol (MGCP) for IPCablecom embedded clients.	IPCable-com Series	2009 (In Review)	Technical			O			

Next Generation 911 (NG911) Standards Identification and Review

Entity	Standard or Document ID	Standard or Document Title	Standard or Document Description	Associated Documents	Latest Revision/ Release Date	Standard or Document Type	Relation to NENA i3 Architecture				
							Client (C)	Access Networks (A)	Origination Networks (O)	ESInets (E)	PSAPs (P)
SCTE	ANSI/SCTE 24-4 2009 (Free)	IPCablecom 1.0 Part 4: Dynamic Quality of Service for the Provision of Real-Time Services over Cable Television Networks Using Data Modems	Describes a dynamic QoS mechanism for the IPCablecom project; facilitates design and field-testing leading to the manufacture and interoperability of conforming hardware and software by multiple vendors.	IPCable-com Series	2009 (In Review)	Technical			O		
SCTE	ANSI/SCTE 24-21 2012 (Free)	BV16 Speech Codec Specification for Voice over IP Applications in Cable Telephony	Contains the description of the BV16 speech codec; gives detailed description of the BV16 encoder and decoder, and contains sufficient details to allow those skilled in the art to implement BV16 encoders and decoders.	IPCable-com Series	2012	Technical			O		
SCTE	ANSI/SCTE 24-22 2013 (Free)	iLBCv2.0 Speech Codec Specification for Voice over IP Applications in Cable Telephony	Contains the description of an algorithm for coding of speech signals sampled at 8 kHz.	IPCable-com Series	2013	Technical			O		

Next Generation 911 (NG911) Standards Identification and Review

Entity	Standard or Document ID	Standard or Document Title	Standard or Document Description	Associated Documents	Latest Revision/ Release Date	Standard or Document Type	Relation to NENA i3 Architecture					
							Client (C)	Access Networks (A)	Origination Networks (O)	ESInets (E)	PSAPs (P)	
SCTE	ANSI/SCTE 24-23 2012 (Free)	BV32 Speech Codec Specification for Voice over IP Applications in Cable Telephony	Contains the description of the BV32 speech codec.	IPCable-com Series	2012	Technical			O			
SCTE	ANSI/SCTE 165-12 2009 (Free)	IPCablecom 1.5 Part 12: PSTN Gateway Call Signaling Protocol	Describes an IPCablecom profile of an API for the Media Gateway Control Interface (MGCI) and a corresponding protocol (MGCP) for controlling VoIP PSTN gateways from external call control elements.	IPCable-com Series	2010 (In Review)	Technical			O			
SCTE	CEA/SCTE J-STD-42-B (Free)	Emergency Alert Messaging for Cable	Defines an Emergency Alert signaling method for use by cable TV systems to signal emergencies to digital receiving devices that are offered for retail sale.	SCTE 18:2012	October 2013	Technical			O			
SCC	ISE I²F (Free)	Information Sharing Environment Information Interoperability Framework (I²F)	Guides the implementation of the ISE information sharing capabilities.		March 2014 Version 0.5	Framework				E	P	

Entity	Standard or Document ID	Standard or Document Title	Standard or Document Description	Associated Documents	Latest Revision/ Release Date	Standard or Document Type	Relation to NENA i3 Architecture				
							Client (C)	Access Networks (A)	Origination Networks (O)	ESInets (E)	PSAPs (P)
SCC	IS&S Playbook (Free)	Information Sharing and Safeguarding (IS&S) Playbook	Aids users in their quest to create or enhance an effective and efficient IS&S environment, and can be used at any point in the environment's lifecycle.		November 2015 BETA Version	Framework				E	P
Telcordia	SR-4163 (Fee/Charge)	E9-1-1 Service Description	Describes the telecommunications network and its associated network elements and features needed to provide E9-1-1 service; describes capabilities of the PSAP that interact with network elements.	Telcordia GR and SR standards	May 1997 Issue 2	Technical Information Document		A	O		
Telcordia	GR-1298 (Fee/Charge)	AINGR: Switching Systems	Provides generic requirements to implement the Advanced Intelligent Network (AIN) technology in a public telephone network.	Telcordia GR and SR standards	November 2004 Issue 10	Technical Information Document		A	O		
Telcordia	FR-EMERG-SVCS-ARCH-01 (Fee/Charge)	Family of Requirements for Emergency Services Network Architecture	Contains the key Telcordia GR documents to support E9-1-1 Service; assists suppliers and network providers in providing E9-1-1 service to their customers in a manner that meets regulatory and customer needs.	Telcordia GR and SR standards	January 2012 Issue 2	Technical Information Document		A	O	E	P

Next Generation 911 (NG911) Standards Identification and Review

Entity	Standard or Document ID	Standard or Document Title	Standard or Document Description	Associated Documents	Latest Revision/ Release Date	Standard or Document Type	Relation to NENA i3 Architecture					
							Client (C)	Access Networks (A)	Origination Networks (O)	ESInets (E)	PSAPs (P)	
Telcordia	FR-NEBS-EQUIP-PROTECT-01 (Fee/Charge)	NEBS™ Family of Requirements for Protecting Network Equipment	Is part of the Telcordia Building Blocks for Network Requirements Integration series; is designed to guide service providers, vendors, manufacturers, etc., in the development and maintenance of reliable communications networks.	Telcordia GR and SR standards	June 2011 Issue 1	Technical Information Document		A	O	E	P	
TIA	TIA-102 Series (Fee/Charge)	Telecommunications, Land Mobile Communications (APCO/Project 25)	Is a collection of 81 documents which define LMR technologies and operational needs.		2015 Edition Includes all current TIA/EIA TSB 102, TIA/EIA-102 AND TIA-102 Standards	Technical Standard (Product/ Design)	C	A	O			
TIA	TIA TSB-102.BACC (Fee/Charge)	Project 25 Interface-RF-Subsystem Interface Overview	Provides an informative overview of key technical aspects and considerations supporting specification of the ISSI.		November 1, 2011 Revision B	Technical Standard	C	A	O	E		

Next Generation 911 (NG911) Standards Identification and Review

Entity	Standard or Document ID	Standard or Document Title	Standard or Document Description	Associated Documents	Latest Revision/ Release Date	Standard or Document Type	Relation to NENA i3 Architecture				
							Client (C)	Access Networks (A)	Origination Networks (O)	ESInets (E)	PSAPs (P)
TIA	TIA TSB-102.BAGA (Fee/Charge)	Project 25 Console Subsystem Interface Overview	Provides detailed specification of the CSSI protocol suite and functional services, establishes the fundamental technical basis underlying development of the other indicated documents.		February 1, 2008 8th Edition Reaffirmation 1/29/2013	Technical Standard	C	A		E	P
TIA	TIA TSB-102.BAJA (Fee/Charge)	Project 25 Location Services Overview	Describes P25 Location Services including a summary of the solution, document suite, and its architecture.		February 1, 2010 Revision A	Technical Standard	C	A		E	P
TIA	TIA-102.BAED (Fee/Charge)	Project 25 Packet Data Logical Link Control Procedures	Specifies the LLC procedures that permit the conveyance of Common Air Interface (CAI) data packets between air interface endpoints for all packet data configurations.		September 26, 2013	Procedural Standard		A			
TIA	TIA TSB-146 (Fee/Charge)	Telecommunications IP Telephony Infrastructures IP Telephony Support for Emergency Calling Service	Covers issues associated with support of ECS from IP Telephony terminals connected to an Enterprise Network (EN); describes new network architecture elements needed to support ECS, and the functionality of those new elements.		March 1 2007, Revision A Reaffirmation November 2012	Technical Standard		A	O		

Entity	Standard or Document ID	Standard or Document Title	Standard or Document Description	Associated Documents	Latest Revision/ Release Date	Standard or Document Type	Relation to NENA i3 Architecture				
							Client (C)	Access Networks (A)	Origination Networks (O)	ESInets (E)	PSAPs (P)
TIA	TIA-222 Revision G (Fee /Charge)	Structural Standard for Antenna Supporting Structures and Antennas	Provides design and performance requirements for communications towers, specifically loading and stability minimums.	TIA-1019; TIA-329.1; CSA-S37; TIA/EIA-411	December 15, 2014 Addendum 4						P
TIA	TIA-568 Set (Fee/Charge)	Commercial Building Telecommunications Cabling Standard Set	Defines structured cabling system standards for commercial buildings, and between buildings in campus environments; defines cabling types, distances, connectors, cable system architectures, cable termination standards and performance characteristics, cable installation requirements and methods of testing installed cable.		September 14, 2015	Technical Standard		A	O	E	P
TIA	TIA-606 Revision B (Fee/Charge)	Administration Standard for Telecommunications Infrastructure	Addresses the administrative needs of a data center as well as that of general administration.		June 22, 2012 Revision B					E	P
TIA	TIA-664.529 (Fee/Charge)	Wireless Features Description: Emergency Services (9-1-1)	Describes services and features so that the manner in which a subscriber may place calls using such features and services may remain reasonably consistent from system to system.	TIA-664 series	October 23, 2007 Revision B Reaffirmation January 30, 2013	Technical Standard (Product/ Design)		A	O		

Entity	Standard or Document ID	Standard or Document Title	Standard or Document Description	Associated Documents	Latest Revision/ Release Date	Standard or Document Type	Relation to NENA i3 Architecture					
							Client (C)	Access Networks (A)	Origination Networks (O)	ESInets (E)	PSAPs (P)	
TIA	TIA-942 Revision A (Fee/Charge)	Telecommunications Infrastructure Standard for Data Centers	Addresses data center design guidelines, structured cabling systems, and network design.		March 2014 Addendum 1	Technical Standard		A	O	E	P	
TIA	TIA-1039 (Fee/Charge)	QoS Signaling for IP QoS Support and Sender Authentication	Provides a QoS signaling standard for use within IPv4 and IPv6 network-layer protocols; adds a security capability which allows sender authentication to greatly increase the network security.		August 1, 2011 Revision A	Technical Standard		A	O	E		
TIA	TIA-1057 (Fee/Charge)	Telecommunications IP Telephony Infrastructure Link Layer Discovery Protocol for Media Endpoint Devices	Defines extensions to the IEEE 802.1AB protocol requirements that support VoIP equipment in IEEE 802-based LAN environments.		April 6, 2006 Reaffirmation August 26, 2011	Technical Standard (Product/ Design)	C	A	O			
TIA	TIA-1191 (Fee/Charge)	Callback to an Emergency Call Origination Stage 1 Requirements	Specifies access network requirements for Callback to an Emergency Call Origination; pertains to 1x Circuit Switched (1xCS) calls routed to a 1xCS access network and 1xCS calls routed to a non-1xCS access network.		August 1, 2011	Technical Standard (Product/ Design)	C	A	O			

Entity	Standard or Document ID	Standard or Document Title	Standard or Document Description	Associated Documents	Latest Revision/ Release Date	Standard or Document Type	Relation to NENA i3 Architecture				
							Client (C)	Access Networks (A)	Origination Networks (O)	ESInets (E)	PSAPs (P)
TIA	TIA-4973.211 (Fee /Charge)	Requirements for the Mission Critical Priority and QoS Control Service	Describes requirements for a mission critical Priority and QoS Control Service for a wireless broadband network; includes requirements to determine a user's default priority on the broadband network, and also provides requirements for dynamic prioritization changes to meet situational needs.		August 1, 2014		C	A	O	E	P
TIA	TIA/EIA/IS-834 (Fee/Charge)	G3G CDMA-DS to ANSI/TIA/EIA-41	Provides general requirements and detailed Upper Layer (Layer 3) signaling radio protocols and procedures for the DS-41 radio interface.		March 1, 2000	Technical Standard		A	O		
TIA	TIA J-STD-110 (Fee /Charge)	Joint ATIS/TIA Native SMS/MMS Text to 9-1-1 Requirements and Architecture Specification Release 2	Define the requirements, architecture, and procedures for text messaging to 9-1-1 emergency services using native CMSP SMS or MMS capabilities for the existing generation and NG9-1-1 PSAPs.	TIA J-STD-110.A, TIA J-STD-110.01	May 1, 2015			A	O		P

Entity	Standard or Document ID	Standard or Document Title	Standard or Document Description	Associated Documents	Latest Revision/ Release Date	Standard or Document Type	Relation to NENA i3 Architecture				
							Client (C)	Access Networks (A)	Origination Networks (O)	ESInets (E)	PSAPs (P)
TIA	TIA J-STD-110.01 (Fee/Charge)	Joint ATIS/TIA Implementation Guideline for J-STD-110, Joint ATIS/TIA Native SMS/MMS Text to 9-1-1 Requirements and Architecture Specification Release 2	Addresses CMSPs and TCC provider deployment considerations of J-STD-110.	J-STD-110	May 1, 2015	Joint Standard		A	O		P
TIA	TIA J-STD-110.A (Fee/Charge)	ATIS/TIA Supplement A to J-STD-110, Joint ATIS/TIA Native SMS to 9-1-1 Requirements & Architecture Specification	Provides errata and clarifications to Joint ATIS/TIA Native SMS to 9-1-1 Requirements and Architecture Specification.	J-STD-110	November 1, 2013	Joint Standard		A	O		P

Appendix B: Standards Gap Analysis

Process	Applicable Standards	Identified Gaps	Gap Addressed in Standards Document?
UE (IMS)	• IETF RFC 6881 3GPP IMS Emergency Services • ATIS focus group on over the top applications • CableLabs	Several are still in development. There is no way to quantify all possible end user devices as related to standards.	**ESIF Issue 74** has been developed and defines an IMS counterpart to the NENA i3 specification.
Access Networks	• 3GPP wireless and broadband IMS networks • Generic IP access networks – IETF RFC 6881 • Cable networks • Legacy selective router • Legacy network gateway • Telecommunications network providers connecting by SS7 or centralized automatic message accounting (CAMA)	IMS networks for OTT origination. Cable networks for both cable specific VoIP and OTT origination, DSL networks for both DSL specific VoIP and OTT origination including possibly FTTC and FTTH. The gap for the legacy selective router gateway (LSRG) was the same as the legacy network gateway (LNG), defining a method for acquiring call related location to enable call routing in NG9-1-1 for legacy wireless calls. This method has been resolved and will be documented in an approved update of the NENA 08-003 (i3) architecture standard. Priority 2.	Call routing partially addressed in **NENA 08-003**, Version 1, page 124. NENA 08-003, Version 2 is in development and could address these gaps.

Next Generation 911 (NG911) Standards Identification and Review

Process	Applicable Standards	Identified Gaps	Gap Addressed in Standards Document?
Origination Networks			
IMS Origination Networks	• 3GPP TS 23.228, 23.167, 24.229 • ATIS IMS ESInet project (P0030)	None	N/A
Non-IMS Origination Networks	• IETF RFC 6881	Possibly cable networks for both cable specific VoIP and Over-the-top (OTT) origination, DSL networks for both DSL specific VoIP and OTT origination including possibly fiber-to-the-cabinet (FTTC) and fiber-to-the-home (FTTH). Priority 2	**RFC 5985 (September 2010)** defines and describes an XML-based protocol that can be used to acquire device location information from an LIS within access networks employing both wired technology (DSL, cable) and wireless technology.
Third-party Originating Service Providers (e.g., OnStar, relay services)	• NENA 08-003 • IETF • TIA	Some are proprietary, but they must comply with ESInet interfaces using a standard public interface. Priority 1	Still needs to be addressed. NENA 08-003, Version 2 is in development and could address these gaps.
Legacy Origination Networks	• Legacy selective router • Legacy network gateway • NENA 08-003 • Telecommunications network providers connecting by SS7 or CAMA	The gap for the LSRG was the same as the LNG, defining a method for acquiring call related location to enable call routing in NG911 for legacy wireless calls. This method has been resolved and will be documented in an approved update of the NENA 08-003 (i3) architecture standard. Priority 1	Call routing partially addressed in **NENA 08-003**, Version 1, page 204. NENA 08-003, Version 2 is in development and could address these gaps.
Femto Cell	• NENA 03-509 v1	Specification needs to be updated for NG911. Priority 3	Still needs to be addressed.

Next Generation 911 (NG911) Standards Identification and Review

Process	Applicable Standards	Identified Gaps	Gap Addressed in Standards Document?
ESInet			
IP network	• NENA 08-003	Testing, Operations Priority 1	Operations partially addressed in **NENA 08-003**, Version 1, page 44. NENA 08-003, Version 2 is in development and could address these gaps.
Core functions (DNS, DHCP)	• IETF	None	N/A
Interconnect with other ESInets	• NENA 08-003	Testing, Operations Priority 1	NENA 08-003, Version 2 is in development and could address these gaps.
Interconnect with origination networks	• NENA 08-003 • IETF RFC 6881	Testing, Operations Priority 1	NENA 08-003, Version 2 is in development and could address these gaps.
Interconnect with access networks	• NENA 08-003 • IETF RFC 6881	Testing, Operations Priority 1	NENA 08-003, Version 2 is in development and could address these gaps.
ESInet to PSAP interface	• NENA 08-003	Testing, Operations Priority 1	Still needs to be addressed. NENA 08-003, Version 2 is in development and could address these gaps.
Interconnection with other emergency service entities	• NENA 08-003 • Other NENA and APCO standards in development	Testing, Operations Priority 1	Still needs to be addressed NENA 08-003, Version 2 is in development and could address these gaps.

Next Generation 911 (NG911) Standards Identification and Review

Process	Applicable Standards	Identified Gaps	Gap Addressed in Standards Document?
Interconnection with other emergency service entities	• NENA 08-003 • Other NENA and APCO standards in development	Testing, Operations Priority 1	Still needs to be addressed. NENA 08-003, Version 2 is in development and could address these gaps.
Management		NENA work in development. Priority 2	Mentions a technical standard is to be determined and will be developed as a guide for PSAP staff and policy makers to evaluate and consider the opportunities and challenges presented with NG911 systems as they relate to personnel and PSAP management.
Location	• 3GPP • ATIS IMS ESInet • IETF • NENA		
PIDF-LO - the location interchange format	• IETF 4119	IMS and IETF/NENA location format incompatibilities. Priority 1	Still needs to be addressed.
Functional definition of LIS (and similar terms)	• NENA 08-003	None	N/A

Next Generation 911 (NG911) Standards Identification and Review

Process	Applicable Standards	Identified Gaps	Gap Addressed in Standards Document?
IP-based Emergency Services	• NENA 08-505	Initial version is incomplete. Future revisions of document are required. Priority 2	**NENA 08-505** (December 2006) acknowledges the first edition of what will be a comprehensive document addressing many access network configurations. This edition has a narrow solutions focus and addresses only the automated mechanism for the residential broadband market.
Location Configuration Protocols		IMS OTT issues. Priority 2	Still needs to be addressed.
Location Dereferencing Protocols	• IETF RFC 6753	Depends on results of ATIS IMS ESInet work. Priority 2	Still needs to be addressed.
Location Query Protocols (to the extent it is decided they are different from location configuration protocols [LCPs])		Pending other work.	N/A
Location Validation	• IETF 5222 • IETF 5223	None	N/A
Interwork to existing location sources, such as automatic location identification (ALI)	• NENA LSRG	None	N/A

Next Generation 911 (NG911) Standards Identification and Review

Process	Applicable Standards	Identified Gaps	Gap Addressed in Standards Document?
GIS & 9-1-1 Attribute Data			
Address, political boundary, and service boundary layer	• NENA GIS V3	None	N/A
Service boundary polygons – how routing occurs	• NENA GIS V3 • NENA 08-003	None	N/A NENA 08-003, Version 2 is in development and could address these gaps.
Data management, quality assurance	• NENA	Further work needed. Priority 2	Still needs to be addressed.
Distribution – how does it get from GIS to everything else	• NENA 08-003 • OGC	OGC work needs further standardization. Priority 1	Still needs to be addressed. NENA 08-003, Version 2 is in development and could address these gaps.
Adjustment of street/address layer to polygon layer	• NENA Emergency Call Routing Function (ECRF)/Location Validation Function (LVF)	Further work needed. Priority 1	Addressed in **NENA 08-003**, Version 1, page 146. NENA 08-003, Version 2 is in development and could address these gaps.
Call Signaling			
Basic SIP call signaling	• IETF 3261 • IETF RFC 6881	None	N/A
IMS SIP call signaling	• 3GPP	IMS ESINET identified some gaps. Priority 1	Still needs to be addressed.

Next Generation 911 (NG911) Standards Identification and Review

Process	Applicable Standards	Identified Gaps	Gap Addressed in Standards Document?
Call Routing			
Routing database (ECRF)	• IETF 5222 • IETF 5223 • NENA 08-003	None	N/A NENA 08-003, Version 2 is in development and could address these gaps.
Routing proxies (Emergency Services Routing Proxy [ESRP])	• IETF 3261 • RFC 6881 • NENA 08-003	None	N/A NENA 08-003, Version 2 is in development and could address these gaps.
Policy-based routing	• NENA 08-003	None	N/A NENA 08-003, Version 2 is in development and could address these gaps.
Media			
Voice	• 3GPP • IETF • NENA	None	N/A
Video	• 3GPP • IETF • NENA	None	N/A
Text	• 3GPP • IETF • NENA	None	N/A

Next Generation 911 (NG911) Standards Identification and Review

Process	Applicable Standards	Identified Gaps	Gap Addressed in Standards Document?
Data only – "non-human initiated"	• 3GPP • IETF • NENA	None	N/A
Real-time Transport (RTT), IMS Multimedia Messaging Emergency Services (MMES), "total conversation"	• 3GPP • IETF • NENA	None	N/A
Accessibility			
EAAC issues and gaps in i3	• FCC EAAC • ATIS INES Incubator • FCC NG911 Notice of Proposed Rulemaking (NPRM)	Identify the teletypewriter (TTY) replacement technology, adoption of that technology, and method of delivering TTY replacement to the NG911 and PSAP. Output of FCC NG911 NPRM may identify additional gaps. Priority 1	Still needs to be addressed. The FCC EAAC Report lists some gaps, and makes recommendations to fill some of these gaps.
Interface between IMS-originating networks and relay services	• FCC EAAC • ATIS	How calls originating from IMS connect to the relay service. Also, given that 911 calls originating on IMS are direct to the ESInet, how do responders get notification that a relay service needs to be involved? Need to have specification developed to define how IMS interfaces with relay services. Priority 1	Still needs to be addressed.
Callback	• 3GPP • IETF • NENA		

Process	Applicable Standards	Identified Gaps	Gap Addressed in Standards Document?
Additional Data about:	• NENA	NENA 71-001: NENA Standard for NG9-1-1 Additional Data – There are significant gaps on how this data is obtained, stored, accessed, secured, and maintained. Priority 1 (generally)	**NENA 71-001** describes the use of additional data available with NG9-1-1 (associated with a call, a location, a caller, and a PSAP) that assists in determining the appropriate call routing and handling. Version 2 will include the EIDD specification.
Call	• NENA 08-003 • NENA 71-001 • IETF additional data • 3GPP • ATIS IMS ESInet	None	N/A NENA 08-003, Version 2 is in development and could address these gaps.
Caller	• NENA 08-003 • NENA 71-001 • ATIS IMS ESInet	Emergency Medical Data Priority 2	Addressed by **NENA 71-001** Appendix A, page 23. **NENA 71-001** describes the use of additional data available with NG9-1-1 (associated with a call, a location, a caller, and a PSAP) that assists in determining the appropriate call routing and handling. Version 2 will include the EIDD specification. NENA 08-003, Version 2 is in development and could address these gaps.

Next Generation 911 (NG911) Standards Identification and Review

Process	Applicable Standards	Identified Gaps	Gap Addressed in Standards Document?
Premise (e.g., floor plans, alarm data, etc.)	• NENA 08-003 • NENA 71-001 • NIST	Further work needed. Priority 3	Partially addressed by **NENA 71-001**, version 1, page 28. NENA 71-001, Version 2, and NENA 08-003, Version 2, are in development and could address these gaps.
PSAP	• APCO • NENA • EIDD	Further NIEM work needed. Priority 1	Still needs to be addressed.
Logging			
Within the ESInet and related functions	• NENA 08-003	NENA and APCO have identified a number of gaps, such as Radio over IP (RoIP). Priority 2	Still needs to be addressed. NENA 08-003, Version 2 is in development and could address these gaps.
Within the PSAP	• NENA NG PSAP	None	N/A
Call origination	• NENA • IETF	Could have IMS and other origination network impacts.	N/A
Bridging/Conference Calls	• NENA • IETF	Could have IMS and other origination network impacts. Priority 2	Still needs to be addressed.

Process	Applicable Standards	Identified Gaps	Gap Addressed in Standards Document?
Security			
Credentials	• 3GPP • IETF • NENA • ATIS IMS ESInet • NIST	Accessibility and privacy controls across the enterprise and diverse systems are still in development.	NIST National Strategy for Trusted Identities in Cyberspace.
Securing protocol Interaction including authentication, integrity protection, privacy	• IETF • NENA 08-003 • ATIS IMS ESInet • NIST	Accessibility and privacy controls across the enterprise and diverse systems are still in development.	NIST National Strategy for Trusted Identities in Cyberspace. NENA 08-003, Version 2 is in development and could address these gaps.
Attack Mitigation	• NENA 08-003 • NIST	None	NENA 08-003, Version 2 is in development and could address these gaps.
End User Location Integrity	• IETF • ATIS IMS ESInet	Standards in development. Priority 3	Still needs to be addressed.
Transition (including data)			
Wireline	• NENA	None	N/A
Wireless	• NENA	None	N/A
VoIP	• NENA	None	N/A
PSAP aspects	• NENA • ATIS RFAI	None	N/A

Next Generation 911 (NG911) Standards Identification and Review

Process	Applicable Standards	Identified Gaps	Gap Addressed in Standards Document?
Relay services (e.g., IP relay, video relay, etc.)	• NENA	None	N/A
TTY	• NENA	None	N/A
Legacy PSAP	• NENA	None	N/A
Testing		Several gaps associated with Testing. Priority 1	**NENA 06-750** is a policy document that reflects changes in: IP technology; implementation and testing; training; and use of building code fire zones to facilitate the creation of the Emergency Response Location.
Self-test	• IETF • NENA	None	N/A
Discrepancy Reporting	• NENA		
Data Management and Maintenance	• NENA	In development.	

www.ingramcontent.com/pod-product-compliance
Lightning Source LLC
Chambersburg PA
CBHW080653190526
45169CB00006B/2097